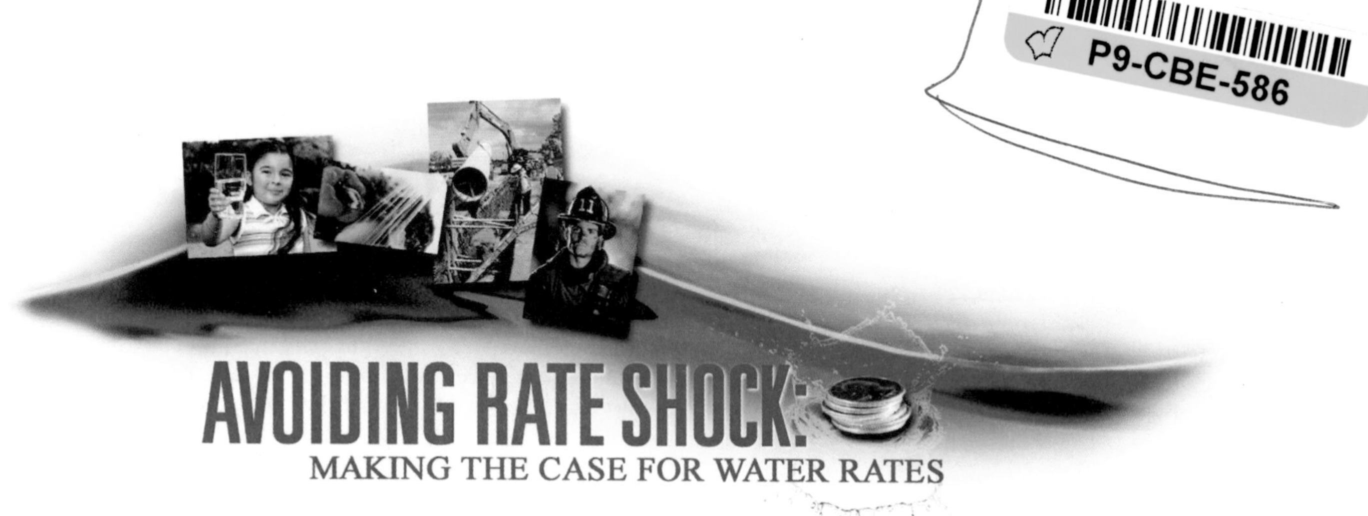

AVOIDING RATE SHOCK:

MAKING THE CASE FOR WATER RATES

A study sponsored by the AWWA Water Utility Council

April 2004

"There is only one thing more painful than learning from experience and that is not learning from experience."

—Archibald McLeish

American Water Works Association

The Authoritative Resource for Safe Drinking Water®

Library of Congress Cataloging-in-Publication Data has been applied for.

Printed in the United States of America

American Water Works Association
6666 West Quincy Avenue
Denver, CO 80235-3098

ISBN 1-58321-334-1

CONTENTS

ACKNOWLEDGMENTS

Workgroup

Joe Gehin, Wausau Water Utility—*Chair*

Bernie Brunwasser, Philadelphia Water Department

Karen Lisowski, Cleveland Division of Water

Ed Olson, Medford Water Commission, OR

Larry Bingaman, Aquarion Water Company

Gary Breaux, East Bay Municipal Utility District

Sue McCormick, Ann Arbor Water Utility

Barbara Buus, Tucson Water

Frank Coulter, Ft. Lauderdale Utilities Department

Project Team

CH2M HILL:

Elisa Speranza—Project Manager

Mike Matichich

Gina Wammock

Vincent J. Ragucci, III

David Green

Xenophon Strategies:

Jay Silverberg

Robert Rosholt

AWWA Staff

Tom Curtis, Deputy Executive Director, Government Affairs

Al Warburton, Legislative Affairs Director—Project Manager

Water Utility Council

Howard Neukrug, Philadelphia Water Department—*Chair*

Julius Ciaccia, Cleveland Division of Water—*Vice Chair*

Mel Aust, Hidden Valley Lake Community Services—*NRWA / Liaison*

Hamlet Barry, Denver Water

Joseph Bella, Passaic Valley Water Commission—*AMWA / Representative*

Walter Bishop, Contra Costa Water District

Michael Burke, NY State Dept. of Health—*Regulatory Agencies Workgroup*

Frank Coulter, Ft. Lauderdale Utilities Department

Paul Demit, CDM—*PAC / Liaison*

Dennis Diemer, East Bay Municipal Utility District

Michael Dimitriou, ITT Sanitaire—*MAC / Liaison*

Terry Gloriod, Illinois-American Water Company

Gregg Grunenfelder, Washington Dept. of Health—*ASDWA / Liaison*

Janet Hansen, Aquarion Water Company of Connecticut—*NAWC / Representative*

Rebecca Head, Washtenaw County Public Health Department—*PIAF / Liaison*

Michael Hooker, Onandaga County Water Authority

Stephen Hubbs, Louisville Water Company—*TAG / Chair (Rep from Other)*

Joseph Jacangelo, MWH—*TEC / Liaison*

Carrie Lewis, Milwaukee Water Works

Barry MacBride, Winnipeg Water and Waste

Kathryn McCain, CDM—*AWWA President-Elect*

Susan McCormick, City of Ann Arbor

L.D. McMullen, Des Moines Water Works

Michael Meadows, Brazos River Authority—*Standards / Liaison*

Edward Olson, Medford Water Commission, OR

Marie Pearthree, Tucson Water

Edward Pokorney, Denver Water Department—*NWRA / Representative*

David Rager, Greater Cincinnati Water Works

John Sullivan, Boston Water & Sewer Commission

Kurt Vause, Anchorage Water & Wastewater Utility

Participating (case study) Utilities

Alexandria Sanitation Authority, VA

Augusta Utilities Department, GA

Cleveland Division of Water, OH

East Bay Municipal Utility District, Oakland, CA

Kenosha Water Utility, WI

Las Virgenes Municipal Water District, CA

Oro Valley Water Utility, AZ

Philadelphia Water Department, PA

Portsmouth Department of Public Utilities, VA

St. Petersburg Water Resources Department, FL

Springfield Water and Sewer Commission, MA

Tucson Water, AZ

Project Funding

Funding for this project was provided by the Water Industry Technical Action Fund (WITAF). WITAF was formed by the American Water Works Association (AWWA) and is funded through AWWA organizational members' dues. WITAF funds activities, information and analysis in support of sound and effective legislation, regulation and drinking water policies and programs.

EXECUTIVE SUMMARY

There is a growing recognition in the United States and elsewhere that the condition of our vital water infrastructure needs serious attention in many places. The pipes, pumps, and treatment facilities that are critical to delivering a clean, adequate supply of potable water to our communities are central to our quality of life, the protection of public health and safety, and a thriving economy. The burden of paying for construction, operation and maintenance of these facilities falls mainly on customers of the systems, and those costs are rising—dramatically in some areas.

The Water Utility Council (WUC) of the American Water Works Association (AWWA)[1] sponsored this study to provide a better understanding of how water utilities might make a good case to their decision-making bodies, customers, and other stakeholders about the need for sustainable local financing of water infrastructure improvements, operations, and maintenance.

The study drew upon background research, case studies, and in-depth interviews with stakeholders, along with lessons learned from the project team's experience working on rate cases and strategic communications campaigns. The team was aided in this effort by a knowledgeable and experienced Workgroup of utility staff from nine geographically and demographically diverse organizations.

> *"Utility credit analysis has moved beyond a point-in-time analysis of current debt service coverage and rates compared to ratepayers' income levels and rates in neighboring communities. Because a variety of factors may affect financing options at the local level, the extent of a utility's ability to implement strategies and policies that address its unique characteristics and allow it to finance needed projects becomes a differentiating factor."*
>
> **—James Wiemken,**
> *Director, Standard & Poors Credit Market Services*

1. AWWA is an international nonprofit scientific and educational society dedicated to the improvement of drinking water quality and supply. Its more than 57,000 members represent the full spectrum of the drinking water community: treatment plant operators and managers, scientists, environmentalists, manufacturers, academicians, regulators, and others who hold genuine interest in water supply and public health. Membership includes more than 4,700 utilities that supply water to roughly 180 million people in North America. The AWWA Water Utility Council (WUC) provides strategic direction and guidance for the legislative and regulatory affairs agenda of the Association.

Findings and recommendations

The key findings of this research will not be news to utilities who have routinely and consistently "done the right thing" by communicating with their stakeholders and keeping up with local funding needs. The findings and recommendations are summarized as follows:

Finding #1. People undervalue water, which compounds the challenge of getting rate increases accepted.

Recommendations:

- Know your customers—understand their attitudes and values through internal and external surveys and other research.
- Clearly explain the benefits of water service and increased spending on infrastructure through simple, graphics-focused presentations, personalized to the neighborhood level.
- Localize messages in alignment with community priorities.
- Participate in local and regional water policy development.

Finding #2. A consistent, structured communications outreach program builds the credibility necessary to support the customer-utility relationship and, therefore, rate increases.

Recommendations:

- Formulate consistent messages, based on an understanding of customer perceptions, and repeat them often.
- Develop and implement an ongoing strategic communications plan.
- Involve employees and decision-makers as "ambassadors."
- Don't make comparisons between water rates and "discretionary" spending (e.g., cable television).
- Celebrate and publicize successful projects, including—and perhaps particularly—underground infrastructure projects.

Finding #3. It's never too late to start doing the right thing—think long-term, and plan beyond the current crisis.

Recommendations:

- Check and re-check your budget and rate calculations.
- Take timing of other events (e.g., elections, increases in other rates or taxes) into consideration and navigate around them.
- Adequately fund and staff communication efforts; assign responsibility and establish accountability.
- Avoid "rate management by crisis" by thinking long-term and starting today.
- Adopt a comprehensive process for planning and implementing a rate increase.

Finding #4. Billing practices and rate structure options can affect customer reactions and acceptance of rate increases.

Recommendations:

- Distinguish water charges clearly from other portions of the bill.
- Communicate and coordinate with any other utilities for which you provide billing, both to (1) control cumulative increases on ratepayers and (2) coordinate communications campaigns.
- Build outreach requirements into combined billing agreements.
- Include enough detail for the average ratepayer to determine how the bill was computed.
- Consider more frequent billing intervals to smooth out impacts on ratepayers and make budgeting easier.
- Consider special rate structures and fee options.
- Use multiple tools and approaches to make sure rates are equitable, defensible, and affordable.

Many AWWA member utilities already are implementing some or all of these recommendations. Many, we know, are not. It is the hope of the WUC and the research team that the findings and recommendations herein will provide useful support and guidance for utilities that have found it difficult to successfully build support for necessary rate increases.

Figure ES-1 is a recommended sequence for planning and implementing a rate increase program, based on a synthesis of best practices and lessons learned from utilities that contributed to this study.

Finally, this report includes examples of tools used by utilities for the successful implementation of rate increases in their communities. The "Avoiding Rate Shock" Toolkit in Appendix B includes examples of brochures, fact sheets, newsletters, presentations, and other materials that utilities can customize to their own local circumstances. For just as "all politics is local," so are all rate increase programs. Paying attention to local circumstances and thinking through the development and delivery of appropriate messages is a theme that echoes throughout the report.

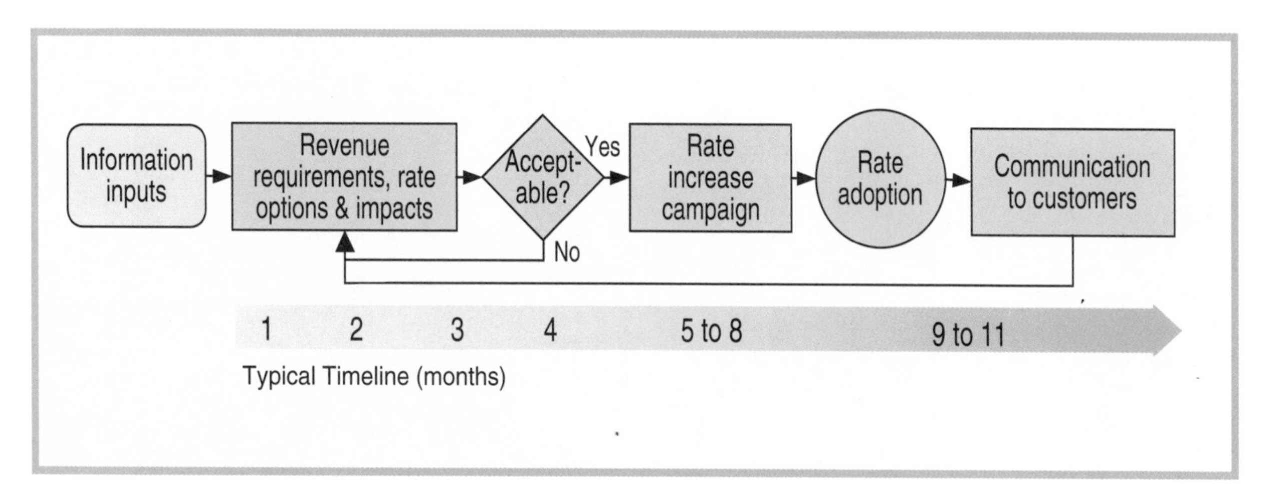

Figure ES-1 Recommended typical rate increase sequence (summary)

A detailed version of this graphic is provided under Finding #3 (page 23) in the body of this report.

INTRODUCTION

In May 2001, AWWA's Water Utility Council published a report entitled *Dawn of the Replacement Era: Reinvesting in Drinking Water Infrastructure*. One of the *Dawn* report's main conclusions was:

> *"Overall, the findings confirm that replacement needs are large and on the way... Ultimately, the rate-paying public will have to finance the replacement of the nation's drinking water infrastructure either through rates or taxes.* ***AWWA expects local funds to cover the great majority of the nation's water infrastructure needs and remains committed to the principle of full-cost recovery through rates."***[2] [emphasis added]

Since that time, numerous reports have been published concerning the projected "gap" between available resources and the needs of local water utilities to address regulatory and infrastructure challenges. Recent reports from EPA and the Congressional Budget Office (CBO)—not to mention competing national priorities and the struggling economy—offer little hope of significant federal assistance any time soon.[3]

Though the needs are indeed great, the local share is not great enough in many places, according to CBO and EPA. They add that federal intervention could create market distortions and disincentives to efficiency, and they note that the lack of an accessible inventory of pipe age and condition compounds uncertainty in projecting costs. In a January 2003 forum titled "Closing the Gap: Innovative Responses for Sustainable Water Infrastructure," EPA highlighted its "Four Pillars of Sustainable Infrastructure," one of which is "full-cost pricing" (the others are better management, efficient water use, and watershed approaches).

According to EPA, *"compared with other developed countries, the United States has the lowest burden for water/wastewater bills when measured as a percentage of household income. Ultimately, prices signal value to consumers and it is important for prices to reflect the increasing*

2. *Dawn of the Replacement Era*, AWWA, May 2001.

3. *Future Investment in Drinking Water and Wastewater Infrastructure*, Congressional Budget Office, November 2002.

scarcity of water. Part of this value includes the increasing financial obligation needed to maintain our water and wastewater systems' infrastructure."[4]

AWWA has echoed some of these same sentiments in its own policy statements on Financing & Rates and Asset Management (please see Appendix C for complete text of the statements). EPA officials have expressed the opinion that water utilities should raise their rates an average of 3 percent per year to help close the gap. Unfortunately, this is easier said than done in many communities, which is why the Water Utility Council undertook this study.

Elected political leaders, authorizing boards, and customers often offer up stiff resistance to funding long-term capital improvements, such as the replacement of aging underground infrastructure, particularly when those improvements are not mandated by court order, administrative consent decrees, or federal law (e.g., Safe Drinking Water Act [SDWA], Surface Water Treatment Rule [SWTR]). Water and sewer projects are generally not perceived as "sexy" projects—unlike, for example, the building of a new school or a park. As one water utility official recently quipped, "When's the last time you went to a ribbon-cutting ceremony for a new pipe?" Therefore, utility managers have employed creative options to secure funding for these much-needed projects, including:

- Increasing rates every year to keep up with inflation
- Planning rate increases to coincide with election off-years
- Adjusting wholesale or commercial/industrial customers to subsidize core city rates and/or residential rates (or vice versa)
- Crafting a strategic communications plan, with stakeholder involvement, to sustain community support for a well-conceived capital program based on documented priorities and needs

While all of these strategies, and others, might be perfectly acceptable means of dealing with the need for additional revenue, the final option on the list was the focus of this project. This report demonstrates the simple premise that aligning a utility communications program with community goals helps build support for rate increases. Typical community goals include:

- Sustainable economic development
- Quality-of-life improvements
- Public health protection
- Environmental stewardship
- Addressing security concerns

By tapping into stakeholders' issues and concerns, utility managers will have a far less frustrating experience in obtaining the rate revenue they need to do their jobs. And if that weren't incentive enough, the credit agencies that issue all-important ratings on utility bonds say that cost recovery through rates is critical to their evaluation of a utility's credit-worthiness.

Project approach

This study is in no way an exhaustive research report on the topic of utility rate increases. No statistically significant surveys were conducted. Instead, the research team attempted to take an existing body of "conventional wisdom" drawn from lessons learned—often the hard way—and present findings and recommendations in an accessible, practical format for water utilities looking for ideas and assistance in meeting their rate increase challenges.

"While a variety of external factors influence this analysis, including regulatory issues, growth trends, customer concentration, and operational capacity, S&P generally looks for rate stability, rate transparency, and long-term planning as relevant factors that are under some control of utility management. Rate-setting procedures that address these issues should help to achieve higher debt ratings, holding other factors constant."

—James Wiemken, Director, Standard & Poors Credit Market Services

4. U.S. EPA, "Sustainable Water Infrastructure for the 21st Century," http://www.epa.gov/water/infrastructure/index.htm.

The Water Utility Council, drawing on a methodology that worked successfully in the development of the *Dawn* report, once again called upon Joe Gehin, General Manager of the Wausau, Wisconsin Water Utility. Mr. Gehin had chaired the "Infrastructure Issues Group" that supported the *Dawn* report, and this time he lent his sage counsel and guidance to the Sustainable Local Financing Workgroup. This Workgroup was formulated to represent a cross-section of system sizes and geographic locations, to get a better sense of the challenges facing different types of water utilities. Some were joint water-wastewater utilities, and one was investor-owned. The Workgroup participants represented a variety of functions within their utilities, including executive management, finance, public affairs, and engineering. (Please see Acknowledgments, p. v, for a complete listing of Workgroup participants.) The group met by conference call and at one in-person workshop to develop and implement the scope of the study.

With the guidance of the Workgroup and AWWA Government Affairs staff, the research team developed two survey instruments intended to elicit relevant data during in-depth interviews: one for the case study participants and one for the stakeholders. Members of the project team conducted interviews and research by phone and Internet, and incorporated the findings from these interviews into case studies. (Please see Appendix A for summaries of the 12 case studies performed.) Figure 1 shows the location of case study utilities.

The stakeholder interviews provided critical perspectives on the findings and recommendations. In the research team's work with water utilities on rates and other issues, it often becomes apparent that there is a gap between what utilities *think* they know about stakeholder attitudes and values and what stakeholders *really* think. (This conclusion was first documented in an AWWA Research Foundation study in 1993.[5]) Just as the stakeholder interviews were an important part of the research supporting this report, they are also a valuable tool for utilities in developing a rate increase program. One cautionary note: utilities should be prepared to both listen to and *heed* the advice and input received from stakeholders. Only through a two-way dialogue will understanding and support be promoted. The range of stakeholders interviewed, mostly in association with particular case studies, is shown below:

Stakeholder type
Council/Board
Mayor
Residential customers
Industrial/institutional customers
Consumer advocates
Business community
Bond rating agencies
Employees

The value of public involvement in general has been well documented by the AWWA Research Foundation and many others. In a 1995 report, AwwaRF said, *"it's possible to have broad support, even in today's skeptical world. The method is to make the public part of the solution rather than part of the problem…to have the utility perceived as service-oriented public servants rather than non-responsive technocrats."*[6]

Drawing upon information gathered from the interviews, the Workgroup calls and workshops, and the experience of the research team, findings and recommendations were formulated to summarize the "lessons learned" and share them with other water utilities. In addition, an "Avoiding Rate Shock" toolkit was assembled to provide templates and examples of communications materials utilities can refer to when developing their own rate campaigns. (Please see Appendix B for this toolkit.)

5. Hurd, Robert, *Consumer Attitude Survey on Water Quality Issues*, AwwaRF Report #90654, 1993.
6. AwwaRF/CH2M HILL, *Public Involvement Strategies: A Manager's Handbook*, 1995.

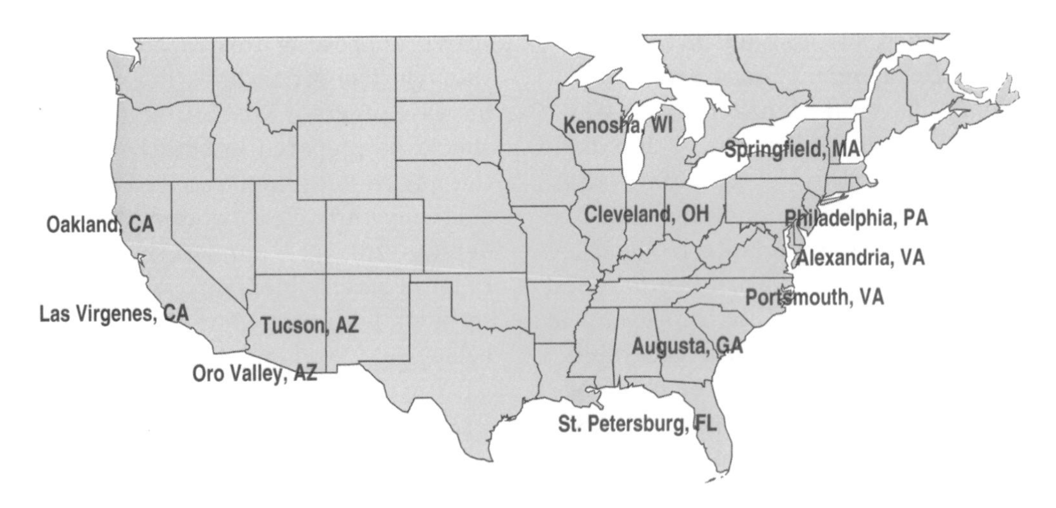

Figure 1 Locations of case study utilities

During the course of researching this report, an important consideration was the impact of rate revenue adjustments on a utility's bond rating. A dialogue with municipal bond rating agency Standard & Poors (S&P) established a number of important factors, which are documented throughout the report. In summary, according to James Wiemken, Director of S&P's Credit Market Services:

> *While a system's current financial status is of some importance to the utility's credit rating, its likely long-term health is the key driver. As the likelihood of significant additional capital needs increases, the current rate, financial, and debt pictures for a utility become less reliable as indicators of long-term credit quality in and of themselves. A utility's ability to implement policies and procedures which garner the support of ratepayers for the additional revenues required to support these needs will become more important to the rating.*[7]

A full copy of the S&P white paper on this topic is included in Appendix D.

In the following sections, we address each of the key findings of this report:

1. **People undervalue water,** which compounds the challenge of getting rate increases accepted.
2. **A consistent, structured communications outreach program builds the credibility necessary to support the customer-utility relationship** and, therefore, rate increases.
3. **It's never too late to start doing the right thing**—think long-term, and plan beyond the current crisis.
4. **Billing practices and rate structure options can affect customer reactions** and acceptance of rate increases.

One of the *Dawn* report's strong recommendations was: "Working with the public to increase awareness of the challenge ahead, assess local rate structures, and adjust rates as necessary… Comprehensive, focused, and strategic communications programs serve the dual function of providing consumers with important information about their water systems and building support for needed investments in infrastructure."[8] The current report provides a framework for water utilities to accomplish this objective.

7. Wiemken, James, Director, Standard & Poors Credit Market Services, "Credit Implications of Rate Structure and Rate Setting for U.S. Municipal Water-Sewer Utilities," white paper, January 20, 2004.

8. *Dawn of the Replacement Era*, p. 22.

FINDING #1:

People undervalue water, which compounds the challenge of getting rate increases accepted.

> ***"The harder the conflict, the more glorious the triumph. What we obtain too cheaply, we esteem too lightly; it is dearness only that gives everything its value."***
>
> ***—Thomas Paine***

Water utilities no longer have the luxury of being the "silent service" they once were. Changing regulations, shifting demographics, rising costs, and the explosion of available information from multiple news sources have forever changed the landscape of public policy.

While right-to-know provisions of the Safe Drinking Water Act Amendments of 1996 have mandated utility communications with customers about the quality of their water, many customers still understand little about how safe water is provided and distributed. Nor do customers in general understand the extent of investment required to continue providing safe water and to keep infrastructure—particularly buried infrastructure—in working condition.

Research shows that most Americans trust their tap water, and when pushed by probing questions in a poll, they will support increased rates and taxes to support safer and more efficient water systems.[9] However, the American public has grown accustomed to paying rates for water service that, in general, do not reflect adequate spending on system rehabilitation and repair.

Decades of under-investing have brought many systems to the point of critical need. Concurrently, there are arguments that water is already approaching the point of being unaffordable in many communities. Further affordability studies are needed to shed light on how much consumers—particularly low-income consumers—can "afford" to pay for water.[10] The vague promise of federal funding and the alternative rate structures used by other utilities (such as electric) may offer solutions to alleviate the burden of rising rates.[11]

The need for increased spending, however, will not await the development of more affordable solutions. Utilities face the challenge today of raising rates to cover near-term spending needs and prepare for sound long-term operation and maintenance.

> *"This is the only country that you can travel on all the compass points to any city you want, turn the spigot on and get a glass of water, drink it and have a very, very high assurance of safe, high quality water. But it is the cheapest natural resource in America for the highest quality in the world. The underpricing of this resource has led to the undervaluing of water."*
>
> ***—Ron Linsky, Ph.D.,***
> *National Water Research Institute,*
> *in address to the Texas Water Summit*
> *Reported in* AgNews, *Texas A&M University,*
> *November 10, 2003*

---

9. A 1999 survey by the Rebuild America Coalition (www.rebuildamerica.org) showed that 74% of Americans were willing to support increased taxes for water projects—higher than support levels for any other public expenditure tested.

10. The Water Utility Council plans to conduct a study in 2004 on issues associated with low income water assistance programs.

11. AWWA/National Consumer Law Center, *Water Affordability Programs*, 1998.

Beyond expected costs, unexpected crises can generate major unbudgeted expenses—many of which may not be optional if a utility is to fulfill its mission of protecting public health. In the August 2003 power blackout that stunned the northeastern and Midwestern U.S., the Cleveland Division of Water's major pumping stations lost power, cutting off drinking water service and water pressure for firefighting. Commissioner of Water Julius Ciaccia said, "We were under the false security that we had plenty of redundancy. Of course, that's all changed now," and the City is "pretty much locked in on off-site generation" for its water system. That translates to an unplanned $20 million to $25 million investment in large diesel generators to ensure the availability of safe drinking water and sufficient water pressure for firefighting in the event of another blackout.[12]

To generate acceptance for the volume of rate increases, utilities must work on changing customer attitudes about the value of water. And to influence customer attitudes, utilities must communicate with customers on *their* terms—which requires understanding how they perceive the value of water today. Customer surveys can provide insight into existing perceptions in individual communities, but some assumptions can be made across the board:

Benefits of increased investment in drinking water infrastructure often are not clearly communicated or understood. To communicate the benefits of increased infrastructure investments, utilities need to educate the public about the need for higher investments—for example, the condition of the utility's assets, the need to enhance security, or the need for data integration to enhance efficiency and improve customer service. Some utilities raise awareness of buried asset conditions by displaying sections of tuberculated water pipelines or other deteriorating system components. In other communities, lack of support for needed rates leads to an informal practice of "waiting until the house is on fire;" publicity related to main breaks and flooding can bolster support for long-needed rate increases. In still other communities, when credibility is in question, even publicity of dire infrastructure conditions and looming regulatory deadlines may not move elected officials and the public to support a rate increase, as in the City of Atlanta, Georgia (Figure 2).

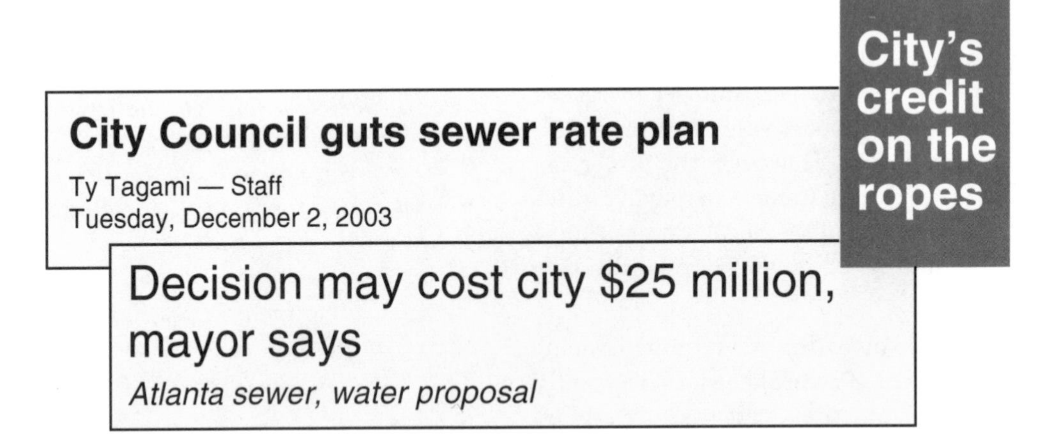

Figure 2 Local headlines on rate increase program—Atlanta, GA

Even extreme regulatory pressure, years of study, an extensive public information campaign, and the threat of poor credit ratings do not guarantee approval of requested rate increases—as demonstrated by the City of Atlanta's experience in 2003.

---

12. *Akron Beacon Journal*, "Averting blackout crisis: Cleveland to invest in generators to keep water flowing in future," November 15, 2003.

Regional variations in the perceived value of water must be considered in communications planning. The perceived value of water in dry and drought-ridden areas understandably is higher than in areas where water supplies are more plentiful. In the latter areas, it can be expected that there will be less public interest in conservation programs, and investments in drinking water infrastructure will more likely be questioned, unless the public has an understanding of the benefits of those investments.

A flashpoint in most American communities is the issue of defining the appropriate level and nature of economic development. Perspectives on this issue vary widely from region to region, and from community to community. Investments in water infrastructure can be seen as encouraging development, and arguments may be made for stronger water conservation measures in the interest of controlling demand and/or reducing costs (most consumers are unaware of the connection between higher conservation, lower revenues, and potentially higher rates). Utilities generally *respond* to demand for infrastructure services, rather than serving as a point of control for development. However, utilities can lose credibility by not participating in the dialogue.

Recommendations

Know your customers. Understand their demographics, their preferred news sources, their political preferences, what they value, what concerns them.

Conduct customer surveys. Conduct surveys and research, both internal and external, including follow-up research. Your customer satisfaction surveys will likely spotlight how your customers feel about the service you provide, and will likely be a reflection of (1) how they feel about the aesthetic quality and perceived safety of their tap water and (2) what, if any, interaction they've had with your utility's meter readers, billing clerks, customer service representatives, and field crews. (Finding #2 includes further information on customer surveying methods.)

Clearly explain the benefits of increased spending on water infrastructure. Supporting data for rate increases might include rate impact projections, multi-year financial models, comparative economic data with neighboring communities, and affordability analyses that compare projected charges with household income and other affordability metrics. *(The Water Utility Council is undertaking a study in 2004 of issues related to low-income rate programs.)*

While financial models and rate impact projections are crucial, they are only a part of the equation. Perhaps the most difficult part of explaining the benefits of increased spending is the task of *communicating* in language stakeholders can understand. Most stakeholders respond to information that is:

- **Presented visually—and *simply*.** If the messages and visual materials you prepare for stakeholder communication cannot be understood in a matter of seconds, you need to simplify them. If you think they're sufficiently simple, they probably aren't. **Try out your messages and visuals** on members of key stakeholder groups, and be on the lookout for negative perceptions or misunderstandings.
- **Personalized**—to a neighborhood level, if practical. The City of Portsmouth, Virginia, built support for a major rate increase in an extremely difficult economy by building its capital plan around neighborhoods—with maps clearly explaining the location and age of current infrastructure (Figure 3).

> *"What do customers want from their water utility in these areas? The same things we all want from our dealings with companies that provide services to us: prompt, courteous, knowledgeable responses and a speedy resolution to our problems."*
>
> —*"Death of the Silent Service: Meeting Customer Expectations," Chapter 17 in* Drinking Water Regulation and Health

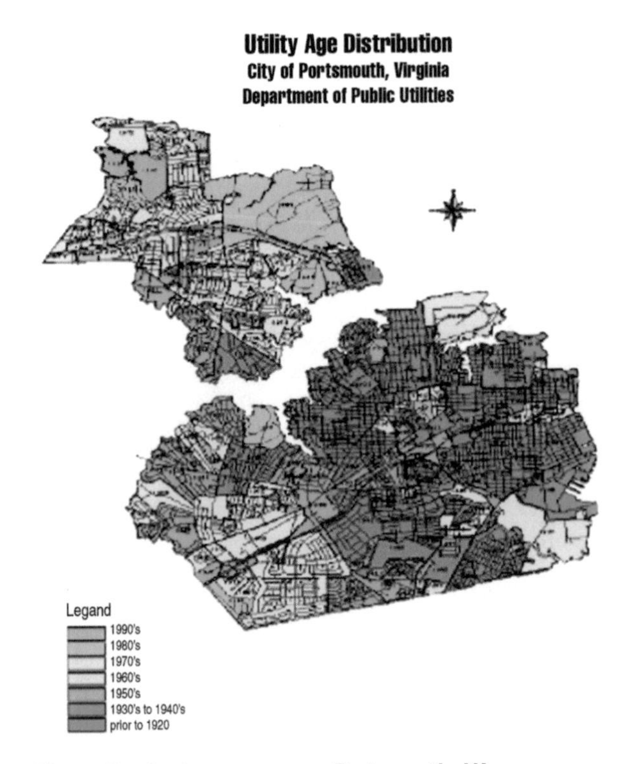

Figure 3 System age map—Portsmouth, VA
In prioritizing neighborhoods for improvement, the Portsmouth, Virginia, Department of Public Utilities used GIS data on average system ages by neighborhood. Operations and maintenance history, such as main breaks and emergency repairs, also were considered. Finally, the Department considered the ability to coordinate the neighborhood system efforts with overall City Council goals for revitalization.

Finding #2 includes detailed information on developing a strategic communications plan for explaining the benefits of increased infrastructure investment (see p. 17).

Localize messages in sync with local and regional priorities. Knowing your stakeholders and their priorities—including the political agendas of elected officials—is just as critical as understanding the realities of water supply and demand issues. Public sentiment around water-related and rate-related issues must be understood and considered in planning communications to support needed rate increases.

Participate in water policy issues. Since utilities often are a focal point when conflicts arise around water issues, generally it is in utilities' interest to take an active role in setting policy around water use and conservation for their communities. The appropriate level of involvement should be defined by the utility's governing board in concert with local elected officials and community leaders.

FINDING #2:

A consistent, structured communications outreach program builds the credibility necessary to support the customer-utility relationship and, therefore, rate increases.

> ***"Facing the press is more difficult than bathing a leper."***
>
> ***—Mother Teresa***

Until recently, it was atypical for many utilities to provide information to customers on the quality of their drinking water. Today, annual Consumer Confidence Reports (CCRs) are required by the right-to-know provisions of the Safe Drinking Water Act Amendments of 1996. Of the 104 million individuals who received CCRs in recent years, 81 million claimed to have read the report.[13] However, individual communities may not echo those results. For example, a recent customer survey in Syracuse, NY, revealed that nearly 75 percent of customers never read the report or did not remember receiving it (Figure 4).[14]

While poll results indicate consumers support increased spending on water projects in general, as cited under Finding #1, it appears there is ground to cover in getting messages across to the rate-paying public.

Effective communication is crucial. The level at which customers will accept news about rate increases can, and many times will, depend on the level at which they have been told to expect their water rates to increase.

An argument can be made that utility customers will judge any news about water—conservation, quality, supply—based on how effectively the news is communicated. The case studies developed for this project support that premise, highlighting the value of consistent, structured outreach to customers about issues related to their water—rates or otherwise. The case studies also demonstrate that outreach efforts can range from modest, to moderate, to costly (e.g., Philadelphia Water Department is required by law to expend an estimated $1 million for each rate case, mainly to fund an intentionally adversarial review process).

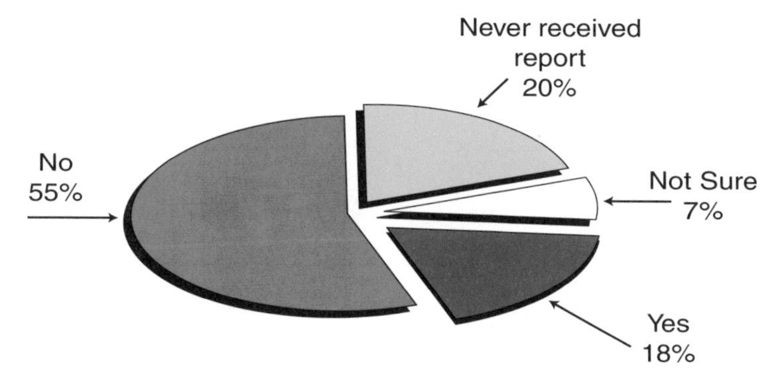

Figure 4 Example of customer polling results—City of Syracuse

13. Analysis and Findings of the Gallup Organization's Drinking Water Customer Satisfaction Survey, U.S. EPA, 2003.
14. City of Syracuse Water Study Summary, Syracuse, NY, 2003. DAPA Research, Boston, MA—Professor David Paleologos.

Yet results were essentially the same in each of the communities that implemented a consistent communications program: rate increases were approved, with limited to virtually no opposition.

As our "Rate Shock Index" (Figure 5) suggests, there is a relationship between stagnant rates/ lackluster communications and regular rate increases/regular communications. The assumptions are these:

- The line measuring "stagnant rates" assumes a utility has not engaged in any type of aggressive rate increase programs, or in active outreach efforts to communicate with its stakeholders.
- The line measuring "regular rate increases" assumes that a utility has engaged in a regular rate increase program, with the commensurate outreach to communicate its goals over the timeline.

Thus, the large red arrow is the "customer shock gap"—representing the effort a utility must undertake to effectively convince its customer base about a rate increase, or possibly to communicate about other issues critical to its operations, when there has been no ongoing outreach. The arrow also is proportional to the time and expense necessary for a utility trying to make its case for higher rates, or any issue affecting its customers.

The smaller, yellow arrow is the "business as usual" indicator for a utility that has a consistent communications program and regular rate increases. These two arrows illustrate the difference in the effort necessary to achieve a needed rate increase between a stagnant rates/ lackluster communications effort and a regular rate increases/regular communications plan.

Consider, for example, the following comments from East Bay Municipal Water District (EBMUD) in Oakland, California, which was facing a multimillion dollar rate increase partly driven by capital improvements related to the 1989 Loma Prieta Earthquake, regulatory mandates, and ongoing capital improvements: "We are in touch with our customers regularly. More importantly, we have had a rate increase program for nearly 20 years—mostly cost-of-living, but at times, a little more. We've always made sure that the increases were explained, and we used various sources—the media, bill stuffers, meetings—and everyone was included... businesses, elected leaders, watchdog organizations," said EBMUD's Gary Breaux,

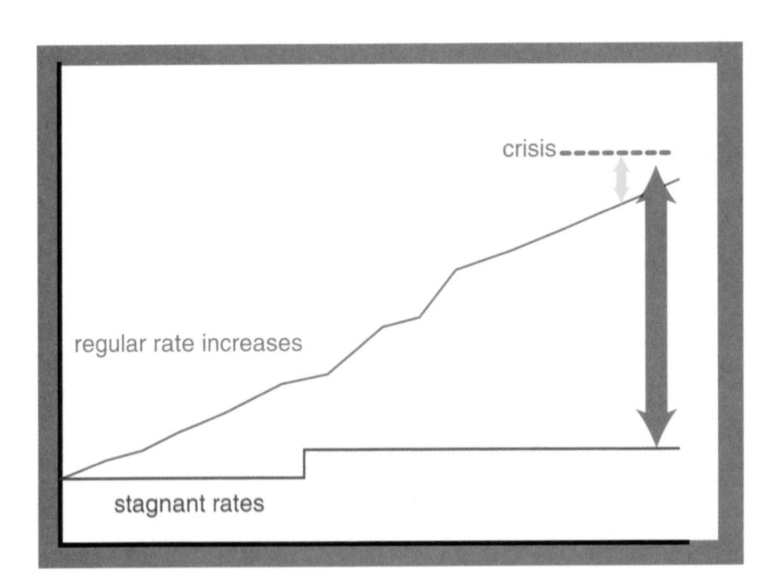

Figure 5 Rate shock index—"the gap"
The choice to manage rates (or other issues) by crisis, or initiate regular increases and community outreach, is linked directly to the amount of time (and/or expense) that a utility will require, and the success it should expect.

during an interview for this project. "We had virtually no opposition to our last increase."

Understanding customer perceptions

How to know what customers are thinking: *ask them.*

A utility that assumes it "knows what's best" for its customers will, in all likelihood, make faulty assumptions about outreach to customers regarding issues affecting them, regardless of the topic—water rate increases, capital improvement projects, bond issues. Unfortunately, many utilities are forced to do just that because they fail to have in place the research means and methods to grasp even a basic understanding of their customers, let alone how those customers receive, or want to receive, information about one of the most important staples in their lives—water.

An August 2003 national study commissioned by the U.S. EPA Office of Groundwater and Drinking Water found the following: the public ranks doctors/public health organizations, state regulators, and environmental groups ahead of water utilities when placing their trust for receiving credible information about tap water (Figure 6).

Further, the EPA study found that the media is the main source of public information about tap water, followed by water utilities and environmental groups. Among the study's conclusions is that *"meeting customer needs, exceeding customer expectations, and providing accurate and timely information on drinking water quality will help develop and maintain public confidence in tap water ...* ***(which) may result in increased public involvement in decision making, and stronger community support."*** [emphasis added][15]

The results of the Gallup poll are echoed in our case studies and the overall experiences of project participants: a utility without at least a baseline understanding of its customers will continually struggle with any type of outreach. By using research to help set the framework for an active communications program, a utility can measure any one, or all, of the following:

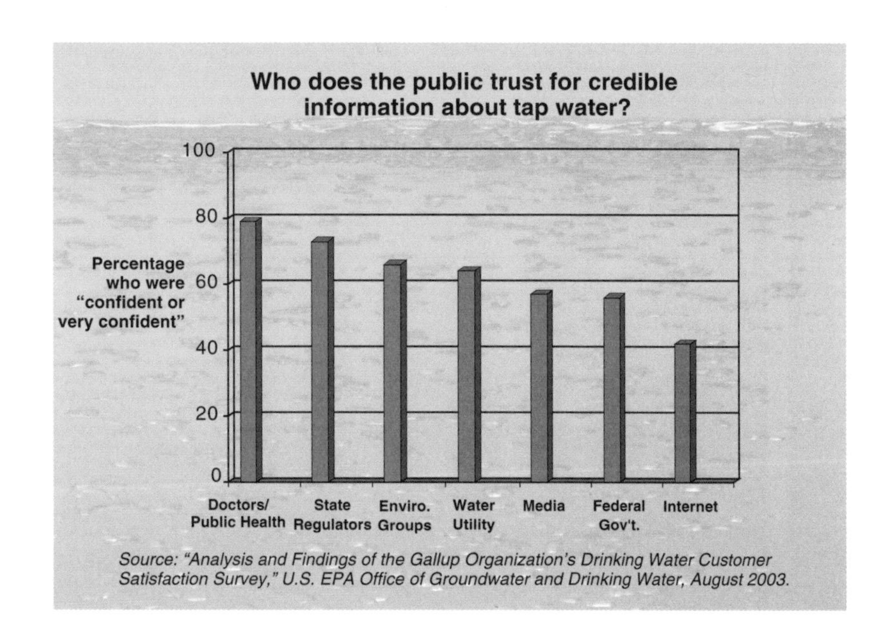

Figure 6 Levels of trust in various sources of information

15. *Analysis and Findings of the Gallup Organization's Drinking Water Customer Satisfaction Survey*, U.S. EPA, 2003.

- **Basic message awareness:** Assuming the utility is attempting to communicate about various issues, are those messages being understood by the targeted stakeholders?
- **New messages:** This is the area where the level of "rate shock" can be measured. Are stakeholders prepared for rates to rise? How much?
- **Spokesperson credibility:** Typical research studies test the public's awareness and trust in certain agencies or communications outlets.[16]

In the course of this project we reviewed several research studies; two in particular support the points raised here. A San Diego County Water Authority research study from January 2003 states, "Not only are the projects, plans, and programs fairly well recognized and well favored, but also there does not seem to be the kind of rate problem that can frequently derail otherwise supported plans and programs." The City of Tampa Water Department's survey in May 2003 showed a high level of customer interest in receiving information about water, and demonstrated effectiveness of Water Department communications about water restrictions.[17]

The Tampa and San Diego projects are examples of the research that can help set the framework for a utility's communications program. Several communities in project case studies cited such research as among the reasons for successful rate increase efforts. The Public Policy Institute of California has conducted studies in partnership with the Orange County Business Council specific to community-wide infrastructure improvements (Figure 7).[18] These types of studies were community-based, involving several hundred people who were contacted randomly by telephone in a fashion similar to more familiar political or brand surveys.

"How important is the condition of the roads and other infrastructure to the quality of life and economic vitality in Orange County?"

		Region		Race /Ethnicity	
	All Adults	North	South	White	Latino
Very important	73%	74%	72%	73%	77%
Somewhat important	25	24	25	25	21
Not important	2	2	3	2	2

Figure 7 Example of customer survey results—Orange County, CA

16. The Pew Internet & American Life Project found in August that nearly half of all Americans are concerned that American utilities such as water, electric and transportation systems, banks and other corporations could be crippled by computer attacks launched by terrorists. This survey found that Americans would turn first to television (57%) and radio (15%) to receive information; 3% would turn to government websites; 9% to government officials [http://www.pewinternet.org].

17. For the full reports, see http://www.sdcwa.org/about/pdf/2003_SurveyReport.pdf and http://www.tampagov.net/dept_water/conservation_education/pdf/Water%20Survey%20Report.pdf.

18. Public Policy Institute of California/University of California, Irvine, *2002 Statewide Survey: December 2002* [http://data.lib.uci.edu/ocs/2002/report/02infrast.html]. The Orange County survey concluded that 78% of the public rated the quality of infrastructure as important to the overall quality of life.

In one case reviewed as part of this project, a telephone survey was conducted solely involving a utility's customers. It delved into current and potential future issues, and it has assisted in building broader support for the utility's programs. Others use mail surveys, such as one conducted last year by Washington State University, to assist in outreach efforts over a four-state region.[19]

Regardless of the methods employed to gain customer and/or public feedback—bill-stuffers, random telephone surveys, brief surveys for customer service operators to use after handling calls—building a baseline of information that provides a window into public attitudes can and will be critical to any utility's communications outreach efforts.

Developing a Strategic Communications Plan

The recommendations accompanying this finding are embodied in the overall suggestion that utilities develop an ongoing strategic communications plan. The guidance provided in this section is applicable to a utility's overall mission, but a rate increase can serve as a catalyst for building a strategic communications plan. Here is a test to determine whether your utility is suffering from *"communications shock"*:

- When was the last time the general manager talked about the utility's overall communications strategy with other senior managers, including the person assigned to oversee the communication function?
- If there is no one assigned to the communications management role, when was the last time the general manager and senior management reviewed the situation to determine whether the organization's interests are best served by maintaining the status quo?
- Have the financial resources assigned to the communications function increased on par with other departments—including salaries for communications staff and materials used for outreach?
- If, within the past 3 to 6 months, the utility was confronted with an emergency response situation, did senior management review the communications response? Was it satisfied that the utility's spokespersons responded appropriately to the situation? If not, did senior management put in place a procedure to make appropriate changes?
- Has the utility surveyed customers to determine their levels of satisfaction with materials they receive from the utility; timeliness of the materials; and trustworthiness of the utility, its management, and its materials?
- Have the results of previous surveys been addressed with changes appropriate to the findings?
- If a television crew or news reporter and photographer walks into the utility's main offices, is the administrative staff prepared to respond appropriately? Are staff in remote or satellite facilities prepared to respond if a member of the media should appear at their doors?
- Have senior managers received sufficient communications training to prepare them for media interviews?
- Did the vulnerability assessment process include discussion about the communications aspects of the various scenarios—e.g., how will the utility communicate should the "what-if" occur; and who will be responsible for communicating?
- When was the utility's website last updated? Does it include such items as utility board decisions; rate explanations; ways for customers to provide feedback about services and/or programs; and contact numbers and email addresses for utility management?

The answers to the test should provide some indication of how close a utility is to "communications shock"—essentially a point at which an issue, a crisis, or a situation has overwhelmed

19. University of Idaho, Survey of Public Attitudes about Water Issues in the Pacific Northwest [http://yosemite.epa.gov/R10/eco-comm.nsf/.

the organization, potentially threatening damage to its reputation, finances, or in some cases, the careers of those in charge. This analysis also might help make the case for adequate staffing and resources to support a public affairs function in a utility.

Informed and involved citizens can either be strong allies of the water utility—or well-armed critics. A comprehensive Strategic Communications Plan can make the difference between creating a constituency and creating a monster. Following are **nine simple steps for developing a Strategic Communications Plan.** If the utility already has a Strategic Communications Plan, it can be compared to and contrasted with the following recommendations.

1. Survey the public and employees

As mentioned above, the case studies support the fact that surveys are used more infrequently in the water business than perhaps in other businesses—yet their value should not be minimized. There are several approaches to using research to help build credibility for communications initiatives.

Survey research

Two of the studies referenced earlier, Tampa and San Diego, were follow-ups to similar research in previous years. Telephone research projects such as these generally cost between $5 and $10 per household surveyed, depending on the number and style of questions asked (multiple choice and/or open-ended) and the time required for the questionnaire. The information can be used in a number of ways: to test stakeholder awareness about utility issues; to test messages that might be used for a utility's initiatives; and to test stakeholder awareness of/sentiment about other issues affecting the community and, thus, the utility.

Focus groups

Telephone research often is paired with focus groups—small, manageable groups comprising individuals selected because they share common interests important to the utility seeking information from them. Focus group members may come from a certain neighborhood, income group, or community interest group. Optimally, a focus group works with a skilled facilitator who is knowledgeable about the issues being discussed, and able to gain more detailed information about people's sentiments than a telephone or website survey might obtain. Total costs can vary but usually range around $5,000 to $7,000 per group.

In-depth stakeholder interviews

Recent research has shown that organizations can get virtually as much relevant feedback from a well-designed and executed in-depth interview as from a focus group. One case study utility (Springfield, MA) used this method to identify themes and messages that were important to various stakeholders; the utility subsequently incorporated those stakeholders' concerns into the rate campaign materials, with great success.

2. Conduct a situation analysis

Organizations engage in issues management on a daily basis. For example, by monitoring new hook-ups, a utility can gauge community-wide growth patterns. The issue: is growth outpacing the utility's ability to supply water? A utility might become aware of a new environmental regulation placing tighter controls over the use of a certain chemical. The issue: how is the utility going to respond? Has it had problems with using this chemical? Are there public records available to the media about the utility's use of this chemical? These are examples of rate increase "show-stoppers"—issues that many times can be anticipated and planned for.

A Strategic Communications Plan should include an active issues management, or situation analysis, process that gives management both a long-term view of the situations it must address, and a snapshot of a situation if management is called on to explain how it is handling an issue.

> *Utilities typically have greater success in raising rates when the increases are introduced as "business as usual" rather than "rates by crisis."*

3. List plan objectives

It is important to take a few minutes at the start of the plan to consider the goals of the communications program. How is success to be measured? Perhaps it's a majority vote on a bond referendum, or a supportive editorial in the local paper. Whatever they are, objectives should be articulated clearly and revisited frequently.

4. Identify stakeholders

In any communications effort, it's important to identify target audiences. Different messages will resonate with people from different perspectives. Identifying who has a "stake" in the utility's actions is an important step. Some of those identified will constitute an audience for the dissemination of public information, and others likely will be more involved in providing direct input to the process. Some of the many audiences and stakeholders the case study utilities identified include:

- Employees
- Local/regional news media
- Environmental advocacy groups
- Mayor/City Manager
- Council or Board
- Multi-family housing property owners
- Regional planning groups
- Bond rating agencies
- State regulators
- EPA
- Public service commissions/public utility commissions
- General customers
- Business community/chambers of commerce
- Industrial/commercial/institutional customers
- Organized labor
- Academic community/universities
- Consumer advocacy groups/fixed income advocates
- Customer advisory boards

Certainly depending on the circumstances, some stakeholders will take on a larger degree of importance and will require more focused outreach. In addition, issues affecting one group will undoubtedly ripple through others. The most important thing is to brainstorm about all obvious and not-so-obvious stakeholders. Many is the utility that regretted forgetting an important constituent, only to have to put out a fire later on. In Tucson, Arizona, for example, the utility conducted extensive outreach efforts to explain the rationale behind a new "system equity fee"—a one-time charge for new users connecting to the water system. Outreach efforts initially appeared successful, but a public hearing revealed that one major interest group had been overlooked: nonprofit developers. Tucson Water worked out a solution with the developers and the City's Community Services Department and the fee was eventually adopted. The system worked because Tucson Water had multiple channels for identifying and addressing stakeholder concerns.

*In **Augusta, Georgia**, commercial and industrial customers warranted special attention. As an early component of its capital program, the utility undertook a major project to implement automated meter reading for its large commercial and industrial customers, who were kept well-informed of increases in charges anticipated to result from the meter program. A bond referendum passed increasing rates by 11% with no organized opposition.*

5. Articulate themes and messages

Messages should be simple, few, and repeated often.

In the case studies developed for this report (Appendix A), messages focused on three themes: maintaining water systems that are (1) safe, (2) reliable, and (3) economically viable. Within those basic themes were supportive messages, such as the need to keep pace with community growth, respond to new regulations, or pay for increasing costs for labor, debt, or infrastructure. Regardless of the venue, the materials used, or the audience, the messages were essentially the same. The theory is no different than a basic premise of political campaigning in which candidates have

A special note on employees:

If a utility can afford to address only one stakeholder group, it should choose its employees. Communicating with customers is the responsibility of all employees—not just the obvious ones like public affairs staff and customer service representatives. Water quality and operations staff, meter readers, the receptionist—all have day-to-day customer contact.

Well-informed employees—whether they receive information from newsletters, intraoffice email, staff meetings, or letters to their homes—can be ambassadors for a utility in potentially hundreds of venues that they might visit: churches, schools, backyard barbecues, athletic events, and others. It is critical to make sure employees are adequately informedso they can accurately represent facts and appropriate messages.

a "message of the day"—the theme they choose to discuss in speeches or media interviews.

For those utilities that managed successful rate increase efforts, the choice seemed clear to them; their communications success was largely within their control. They followed a basic tenet for any communications outreach: *control the message, or someone else will.* Here are some examples of messages that worked well for some of the case study utilities:

> ***"Infrastructure investments support our quality of life and provide a legacy for future generations."***
>
> ***"We are committed to improving efficiency—we're doing our part to control costs."***
>
> ***"Our rates have not kept up with inflation."***
>
> ***"Paying for renewal and replacement is less costly in the long run."***
>
> ***"Failing infrastructure is bad news for our economy."***
>
> ***"A water and sewer infrastructure replacement program will increase the reliability of our underground life support systems."***
>
> ***"A sound water and wastewater system is essential to economic development."***

The case studies also provided different approaches to making a point, such as:

- Rate impact projections [*message:* economic viability]
- Comparisons with neighboring communities' water rates [*message:* fair and equitable increases]
- Demonstrating a connection between sound utility assets and meeting other community goals, such as economic development [*message:* reliability]
- Analyses comparing projected increases to household incomes [*message:* fair and equitable increases]
- Maps showing the age of pipes in various neighborhoods [*message:* reliability, safety]
- Visual aids and facility tours to illustrate the condition of existing infrastructure [*message:* reliability, safety] (an example of this technique is described in Figure 8)
- Comparisons with rate increases in other utilities (e.g., electric, gas) [*message:* fairness, reliability]

Important note: Utilities have found that making comparisons between water rates and "discretionary" spending (e.g., cable TV) can backfire. Appendix B has examples of recommended and *not recommended* ways to present comparison messages.

While the messages case study utilities used were similar in theme and content, the avenues for delivering the message varied. In some cases, the utilities made certain that their messages were in line with regional water or environmental priorities. Some methods, of course, are required by law; others are common communications practices.

6. Use a variety of communication tools

The case study utilities used a range of methods to communicate about rate increases:

- Public hearings

Figure 8 Demonstrating the need for infrastructure improvements

In support of its rate increase, the Cleveland Division of Water used visuals to demonstrate the condition of the water system, including maps, pictures and samples of tuberculated pipes, as well as field trips for Council members to see first-hand the condition of its treatment plants. In January 2000, the community got a live demonstration of the need for infrastructure rehabilitation when a 75-year-old pipe failed, flooding city streets and disrupting water service in the heart of downtown Cleveland.

- Presentations to Councils/Boards
- News releases
- Websites
- Mailers
- Public information meetings
- Brochures
- System tours
- Media briefings and/or editorial board meetings
- Newsletters
- Fact sheets
- Public displays

Appendix B, the "Avoiding Rate Shock" Toolkit, contains some examples utilities can modify to fit their own local circumstances and needs.

7. Develop an action plan

Regardless of the stakeholder group or setting, preparation is critical. The advance work has to anticipate the materials that will be used; how they will be used; who will present them; who will serve as spokesperson, or, if there are multiple spokespersons, who will speak when; how questions will be handled; what support staff will be required (the general manager should not be expected to know the answers to everything!); what arrangements should be made for media coverage; and what community activists might attend, and what are appropriate responses for their issues. Sometimes a simple matrix detailing who, what, when, and resources needed can help make implementation more straightforward.

Follow-up is equally crucial. Once an issue has been introduced to stakeholders, the flow of information needs to continue, if not to keep stakeholders abreast of developments, then to validate the original intent of the initiative. Months after a rate increase is approved, as funds received are used, stakeholders should be informed about the progress being made.

8. Implement and monitor the plan

Once the plan is in place and those charged with carrying it out are all "singing from the same sheet of music," it is time for implementation. It is important to monitor progress, making midcourse corrections (perhaps adding in

Figure 9 Publicizing and celebrating successes
A neighborhood-based infrastructure program in Portsmouth, VA, supported successful implementation of a rate increase—and an achievement award from the Virginia Municipal League, which stated, "… while many communities struggle to find the money to replace unseen infrastructure, Portsmouth has shown that this obstacle can be overcome—even in the most difficult circumstances."

some attention to that forgotten stakeholder) along the way.

Once objectives have been accomplished, **don't be afraid to publicize successes**. The case study from Portsmouth, Virginia (Figure 9), is a good example of an infrastructure program that gained public support despite difficult economic times, and won a state-wide award in the process.

9. Incorporate regular feedback

Talking to customers only about rate increases may be viewed by some as proactive communications, but if rate increase information is all they are receiving, it is doubtful customers will be eager to pay more. However, a utility with an active communications campaign to keep customers informed about infrastructure improvements, water conservation, cost reductions, water quality improvements, and environmental protection will probably be more likely to gain the support of customers and other stakeholders. Regular survey research or other customer and employee feedback mechanisms will help fine tune plans to make them even better the next time.

FINDING #3:

It's never too late to start doing the right thing—think long-term, and plan beyond the current crisis.

> ***"A journey of a thousand miles begins with a single step."***
> ***—Chinese proverb***

Some utilities are a bit overwhelmed by the notion of a rate increase, and unsure of how and where to begin planning for them. Figure 10 (p. 27) depicts a recommended sequence for implementing rate increases, along with a typical timeline. A detailed narrative description of the process is included later in this section.

The sequence shown in Figure 10 is intended as a continuous process. Where to begin along this continuum depends on each utility's unique situation. Is there some sort of crisis or triggering event? Are there political considerations (are there ever *not* political considerations)? Is there resistance from a particular stakeholder or group? These and other factors must be taken into account in deciding where to plug in; the important thing is to *do something*.

Success in implementing a rate increase will depend largely on how successfully the utility has managed other factors, communications included. Rare is the crisis, negative news story, contentious public hearing—or all of them balled into the same situation—that doesn't come with some early warning. Consider these questions:

- Has your incidence of main breaks and other infrastructure failures increased in recent years?
- Have you delayed rate increases for more than five years?
- Have infrastructure failures led to negative media coverage?
- Have you reduced a rate increase below what you really needed for fear of rejection?
- Are you operating under a consent order to expand or improve your system?
- Have your neighboring communities all raised rates while you have not?
- Are you planning a system of upgrades to comply with new regulations or security requirements?
- Have your rate increases failed to keep up with inflation?
- Have you been unable to meet customer demands?
- Have you failed to update your rate model in the past two years?

Issues such as these are like cracks in a foundation. If noticed quickly and repaired, the damage can be contained and the mitigating reasons addressed *before* there is a crisis. If not, public opinion toward efforts to address the problems can be negatively impacted when a crisis does arise. Communications around a rate increase can be difficult enough; a defensive position is the worst place to start.

If utility management fails to manage "showstoppers"—essentially those issues or situations that can derail a rate increase—any communications outreach will become exceedingly difficult. The credibility of any message will be devalued, and the ability of utility management to effectively respond to critics will be undermined.

> *"When managed from a long-term perspective, sound policies usually benefit both bondholders and ratepayers, and the interests of these two constituencies are more consistently aligned."*
> ***—James Wiemken***
> ***Director, Standard & Poors Credit Market Services***

Recommendations

Get the numbers right. Check and re-check budget and rate calculations. One of the project participants, a former newspaper editor, remembers a public official meeting with his newspaper just weeks before a tax increase vote to finance local infrastructure improvements. During the meeting, the official's chief of staff interrupted and placed before his boss "new numbers" that countered what the official had been discussing for nearly an hour. The measure didn't get the newspaper's endorsement, and it failed overwhelmingly. *Lesson:* If the numbers can't be supported, no amount of communicating will make them work.

In Portsmouth, Virginia, the Department of Public Utilities developed a number of related technical studies to support the rate increase effort, including a cost-of-service report, rate impact projections, multi-year financial models, rate comparisons with neighboring communities, asset management and community goals, economic development and affordability analyses. Given the community's large low-income population and the limited amount of taxable property, these studies were instrumental to the rate increase campaign's success.

In the late 1980s, ***Orange County (Florida) Utilities*** *(OCU) adopted an automatic annual increase for water and wastewater rates. This came after much difficulty was encountered in implementing substantial required rate increases in the mid-1980s. OCU's Board adopted a resolution stating that on October 1 of each year (start of fiscal year), water and wastewater rates would increase by 3% as long as no other action was taken. Each October bill included the new rate with a statement reminding the customer of the "automatic" increase. After a few years, the utility's revenues were sufficient. In some years (usually politically timed), the Director was able to go to the Board with a request to formally waive the automatic increase for the coming year. Of course, this was greeted with great enthusiasm, and it fostered much goodwill with customers. By creating the expectation that normal, inflationary increases should be expected, the utility was able to keep up with rising costs and program needs.*

Timing is everything. Take timing of other events (e.g., elections) into consideration and navigate around them. Another example of a "show-stopper" might be when a rate increase effort becomes intertwined with another public vote or election—with the potential for the issues to be confused, the messages diffused, and the electorate unable to clearly understand (or support) any of them. Even the best of intentions can be lost amid a ballot full of tax increase issues. Granted that legislative or other regulatory concerns may determine the timing for rate increase initiatives, but a clear path to customers and/or voters generally is easier than having to debate the merits of a utility increase versus tax funding for other infrastructure needs. Timing of rate increases, as several case studies showed, can be critical to the success or failure of such efforts.

Make communication somebody's job. Adequately fund and staff communications efforts; assign responsibility and establish accountability. Most utilities have staff in place to determine budget needs, conduct rate analyses, and make timing decisions. The case study subjects also have staff assigned full-time to communications. Position titles may be "communications director" or "community affairs director" or "public affairs director"; the responsibilities and roles are virtually the same. Yet, as we know from previous projects and from research on this one, the utility industry is no different than others in that the communications function can vary—from one of critical importance with access to senior management and a key role in all organizational outreach, to simply a functionary with limited responsibilities or resources, to the communications role folded into the responsibilities of a mid-level or senior-level position.

We can say from experience that **the success of an organization's communications program generally parallels the importance management places on the function,** both

*In **Springfield, Massachusetts**, a good strategy and clear, effective communication materials were just the tools. The real success lay in the Commission's ability to draw upon a "reservoir" of credibility with the community—personified by an experienced, well-respected Executive Director and a full-time, high-level staff person in charge of communications. They told the truth, answered questions in a straightforward manner, and were not afraid to say "we have some challenges here, but we're working hard to meet them."*

in terms of communications leadership, access to organizational management, resources to meet its goals, and overall vision for the position. Some 'rate shock' is probably inevitable—it's unlikely rate increases will ever be met with warm receptions. The severity of the response to rate increases, however, is often directly related to the level of attention paid to communications over a sustained period of time.

Avoid "rate management by crisis." Think long-term and start today. According to Standard & Poors:

> *"Many utility officials site the impossibility of correctly estimating future economic development trends, regulatory outcomes, and the long-term patterns of various cost pressures. As such, they claim that trying to measure them actually represents a poor use of limited resources, especially for smaller systems that lack the staff or funds for consultants to devote to such studies. While most of these drivers are indeed highly uncertain, **Standard & Poors views a refusal to consider the potential burden of pressures beyond the short to medium term to be a credit risk.** Accordingly, even small utilities that have attempted to examine long-term risks and possibilities in limited ways consistent with their resources and capabilities will likely find their rate projections and capital plans more accepted by S&P."*[20] [emphasis added]

In Alexandria, Virginia, careful planning, which documented the need for—and affordability of—the program, helped the Alexandria Sanitation Authority (ASA) qualify for substantial amounts of low-interest loans, which have lessened the program's impact on ASA's customers. To date, the program has qualified for more than $99 million in low-interest state revolving fund (SRF) loans, and the advance rate and financial planning have been key elements in securing this funding.

If there is no plan in place, begin today. Change the customer service script; review the website to determine if it is as up-to-date as necessary. Include communications strategy as part of senior management meetings. As for the cracks in the foundation, they can be filled now for less than it will cost if the foundation gives way and the basement begins to flood.

Adopt a comprehensive process for planning and implementing a rate increase. The schedule of activities for securing support for necessary rate increases can vary widely. Several factors can affect the steps and time required for each stage—for example:

- **Approval process**—If a rate hearing and/or public utilities commission (PUC) filing is required, that can add a month or more to the required schedule.
- **Magnitude of increase required**—More planning and outreach may be required for a large increase.
- **Resource constraints**—If a utility has limited internal outreach resources, additional time may be required to secure help for the public information phase.

Recommended sequence for rate increase planning

Despite the variables, it is possible to identify a typical sequence of events to optimize the chance of success in getting rate increases approved

20. Wiemken, James, Director, Standard & Poors Credit Market Services, "Credit Implications of Rate Structure and Rate Setting for U.S. Municipal Water-Sewer Utilities;" white paper January 20, 2004.

(Figure 10 on page 27). This sequence can be adapted to each utility's specific context.

Step 1: Determine annual revenue requirements

The first step in the process is to figure out how much revenue is required beyond the revenues that are produced by existing rates. As shown in Figure 10, the amount of revenues required are typically determined from several sources, which may include:

- Input from customer advisory boards
- Customer surveys (to determine appropriate service levels)
- Capital Improvement Plans, Facilities and Compliance Plans
- Operating budgets, which identify the requirements for labor, materials, debt service payments for borrowed funds, etc.

Some utilities plan for and adopt rates for multiple years in a single action of their governing boards. Even when the rates are being adopted for only a year at a time, it is strongly recommended that the revenue requirement and rate forecasts be developed for a planning horizon of at least 5 years. Taking a long-term view in planning for revenue requirements and rates helps reduce the likelihood of a massive rate spike in any single year.

Step 2: Identify financing/rate options worth studying, and identify affected stakeholders

Step 2a: Identify range of financing, rate and fee options worth considering

Based on the magnitude of revenue requirements identified, the financial management team[21] is positioned to identify the specific scope of the required financial studies. For example, if the revenue requirement is only a few percentage points above revenues from existing rates, the team may conclude that funding options can be limited to flat percentage increases above existing rates. If the revenue requirements are substantially greater than revenues from existing rates, the team may conclude that there is merit in exploring a number of funding/rate structure options, such as reclassified customer groups or increased use of system development charge. Appendix B identifies some of the rate and financing options that should be considered in the latter case.

Step 2b: Identify or confirm key affected stakeholder groups

It is important to identify the stakeholder groups that will be most strongly affected by the changes in revenue requirements identified in Step 1. In many cases, utilities already will have identified their key stakeholder groups during prior rate increase efforts, customer surveys, or other outreach efforts. Even so, it is important to revisit the list and to make any adjustments needed in light of the size of rate increases and potential changes to the rate structure. If a very large increase in revenues is required, almost all customers likely will experience significant rate increases. If a utility decides to evaluate rate options that are more targeted, it is important to identify the affected groups. For example:

- If it makes sense to explore adding/expanding a system development charge, the development community likely will be more heavily affected than general customers. Specific communications to this group should be planned.
- If a utility is considering reclassification of charges to assign more of the costs to customers that contribute to system peaking requirements, the rate increases will particularly affect commercial and industrial customers who have peaking demands that are hard to manage/move.

---

21. The term "financial management team" is used here to refer to the team of utility staff members who will be involved as the core group in developing and implementing the rate increase campaign. In smaller utilities, this team may be limited to two or three people (the finance director, utility manager, and public affairs officer). In very large utilities, this team may consist of half a dozen or more additional people, such as people responsible for developing the utility's Capital Improvement Plan, someone responsible for oversight of the utility's rate model, representatives from the utility's engineering department, and others.

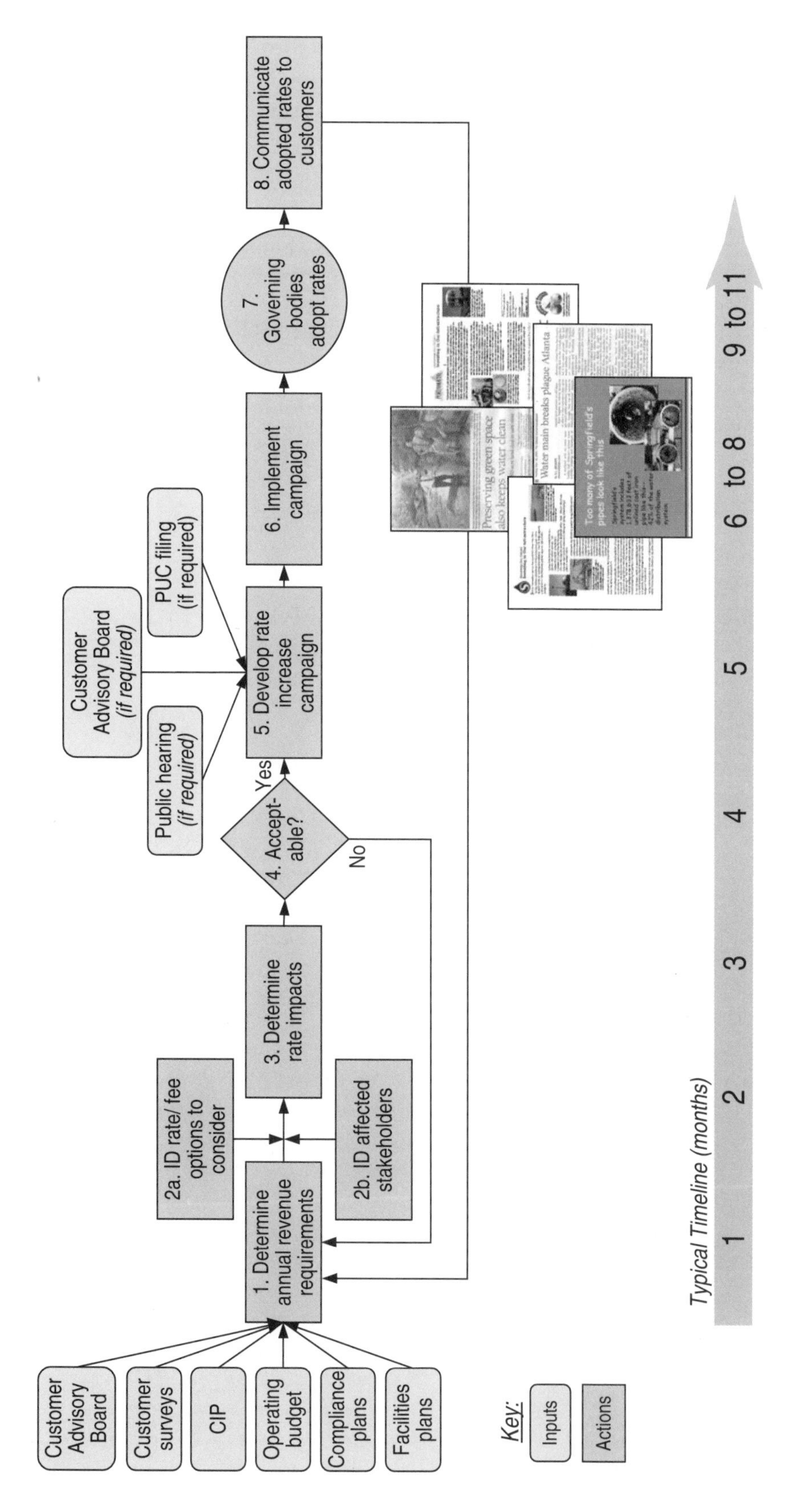

Figure 10 Recommended typical rate increase sequence

The affected parties identified at this point will be targeted in communications plans as the rate increase outreach campaign is developed in Step 5.

Step 3: Determine rate impacts

Based on the revenue requirements and rate/financing options identified in Steps 1 and 2, the utility should identify the rate impacts. This often involves updating or revising the utility's rate model. Key outputs of this analysis will include the required rates and resulting charges to typical residential, commercial, and industrial customers. These results can then be compared to current rates/charges, rates in neighboring communities, and suggested affordability guidelines.

Step 4: Determine acceptability of rate proposal

From the analyses conducted in Step 3, the financial management team should determine whether the rate impacts are likely to be opposed by the utility's stakeholders and governing body. Staff should brief the governing body first on preliminary results and to develop the outreach campaign (Step 5). In some cases, it may be appropriate to brief relevant committee/Board chairs or particular members of the governing body who are more likely to have an interest or questions, or who represent a particular constituency. Many utilities also have advisory groups or particular stakeholders with whom they can "test the waters" before proceeding with the next steps of the rate increase process. If rate increases are *not* acceptable, the utility must revisit Steps 1 and 2 to consider options for multi-year rate increases, reducing the revenue requirements, and/or evaluating additional rate structure/fee/financing options.

Step 5: Develop the rate increase campaign

In this step, details of the outreach campaign are developed. When developing the specific program, the financial management team should consider factors such as:

- Whether a public hearing is required
- Whether a referendum vote is required to enable identified borrowing requirements
- Whether a PUC filing is required
- Whether input is required from a Customer Advisory Board
- How many meetings/briefings should be conducted with specific customers/groups
- What media are available, and which have been successful in earlier rate increase campaigns or other outreach efforts

The Strategic Communication Plan described in Finding #2 identifies in more detail the elements that should be considered when developing a rate increase campaign.

Step 6: Implement the campaign

Depending on the action plan identified in Step 5, this step might include activities such as:

- Public information meetings
- Briefings of individual Board members or committees
- Public hearings
- Focus groups
- Billing stuffers
- Website postings
- Press releases
- Cable access television programs

Step 7: Governing bodies adopt rates

The culmination of the process, optimally, is successful adoption of rates by the utility's governing body or bodies. Depending on the context, this might be an action by a City Council or County Board of Supervisors—or it might be a vote by a separate governing board for an independent regional water authority. For regulated

*The **Springfield (Massachusetts) Water and Sewer Commission** raised rates in support of a $70 million capital program of infrastructure improvements. Without investing in its underground infrastructure, the City anticipated that disruption from almost weekly main breaks would only have become worse. "I believe that the presentation materials and advance preparation of the capital plan helped to educate the public on the need for the improvements and the rate increase to support that work," said Springfield Water and Sewer Commission Executive Director Joe Superneau.*

rate contexts, this also might include approval of the rate increase by a state PUC or Public Service Commission. In addition to briefing Board members, the media, and any other relevant stakeholders in advance of the meeting, one or more representatives of the financial management team should be briefed on key issues so they are prepared to answer any questions raised by governing bodies at the adoption meeting(s).

Step 8: Communicate adopted rates to customers

As stated earlier, rate adoption does not end a utility's communication responsibilities for a successful rate program. Particularly if there is a significant rate increase or a major structural change to rates, it is critical to follow up with additional communications during rate increase implementation. This can include such activities as:

- Distributing follow-up brochures or bill-stuffers
- Posting information related to the implementation of the new rates on the utility's website
- Updating listings of frequently asked questions (FAQs) in response to questions asked during implementation—and providing training on responding to those questions for customer service representatives, meter readers and others with customer contact
- Operating customer service phone banks to respond to questions as bills are sent out

Information gained during implementation provides additional inputs into the planning for the utility's next rate increase.

Range of timeframes

As indicated earlier under this finding, the timeframes required to secure the recommended rate increases can vary widely, depending on the size of the rate increase, the size and complexity of the utility and its service area, and other factors. As revealed by the sample range of timeframes described below and illustrated in Figure 10, early planning is desirable—as a successful, well-organized campaign occurs over a matter of months, not weeks:

- **Step 1:** Determine annual revenue requirements—Allow 1 month from the time that the identified input sources (e.g., surveys, capital planning documents) are available.
- **Step 2a:** Identify financing, rate and fee options—Allow 2 to 4 weeks from the time that the overall revenue requirements are known.
- **Step 2b:** Identify or confirm key affected stakeholder groups—Allow 2 to 4 weeks from the time that the overall revenue requirements and range of finance and rate options have been identified.
- **Step 3:** Determine rate impacts—Allow 1 to 2 months from the time Step 2 has been completed, depending on the complexity of the rate structure, amount of time since last rate model update, and number of scenarios that need to be evaluated.
- **Step 4:** Determine acceptability of proposed increase—This step should be accomplished quickly, before details of the proposed increase are inaccurately reported; allow about 2 weeks.
- **Step 5:** Develop rate increase campaign—Allow 1 month from the time Step 4 has been completed.
- **Step 6:** Implement campaign—Allow 1 to 3 months, depending on the complexity of the issues, number of meetings and brochures that are required, etc.
- **Step 7:** Governing bodies adopt rates—Governing bodies operate at different paces; in general, at least 1 month should be allowed from the end of the communication campaign for the governing bodies to act.

As shown in the timeline at the bottom of Figure 10, these activities result in a typical schedule of 9 to 11 months. This reinforces the importance of planning ahead and avoiding the too-frequent occurrence of trying to compress these activities into a 3- to 4-month period of time.

The probability of securing rate increase approval from governing bodies and support from customers is much higher when adequate calendar time and resources are devoted to the effort.

FINDING #4:

Billing practices and rate structure options can affect customer reactions and acceptance of rate increases.

"What people don't understand, they will not value; what they don't value, they will not support."

—Andy Richardson

The timing and format through which water utilities bill their customers can have a major influence on how rates are perceived—and, therefore, on how readily rate increases are accepted. The rate structure used, and the way it is explained to customers, also can significantly influence the acceptability of rate increases by customers at large, or by sub-groups within a utility's customer base (e.g., low-income or elderly customers, large-volume industrial customers). The discussion below identifies some key issues related to billing practices and rate structures, and it recommends ways to use these mechanisms to encourage customer acceptance of rate increases.

Billing practices and format

Several characteristics of water utility billing can raise issues that influence the way retail customers understand and perceive their water bills.

1. Stand-alone bill vs. multi-service bill

Some water utilities send out bills that include only the charge for water service. However, in many cases, the water bill also is the billing vehicle for a variety of other charges, such as wastewater utility charges, stormwater utility charges, or solid waste charges. Usually it is cost-efficient to group such charges on a single bill, reducing the need for multiple billing databases and other redundant charges, such as postage costs, and reduces the number of checks the customer has to write for utility services.

In most cases, the water utility has become the most convenient vehicle for delivery of such multi-service bills, because typically the water utility has a broader customer base and coverage area than the other services (and water utilities have the only tangible, metered connection to the customer). However, multi-service billing raises issues related to customer perceptions. For example, for most municipal utility systems, there are more water-only customers than there are sewer-only customers. Because the bill typically comes from the water utility, it is often viewed as the entity imposing all the billed charges on the customer. This poses several challenges to the water utility, including:

- **Customers don't differentiate.** Where the charges for sewer, waste disposal, streetlights and other services are included on the same bill, **often the consumer doesn't differentiate between them**. Sewer bills may increase astronomically while water rates are relatively static, and the **perception likely will be that water rates are rising** alongside others, particularly because the customer's bill comes on water utility letterhead.[22] For example, Tucson Water also does billing for the county's sewer system. Figure 11 illustrates a sample bill. While the Tucson water system is on a rate plan that is based on gradual increases, higher

22. There are some exceptions to this practice. For example, in the case of Aquarion, an investor-owned water utility, sewer bills are based on water bills, but they are billed separately.

increases in sewer rates would be expected to affect perceptions of water rates as well.

- **Aggregated bills increase affordability and solvency issues.** Retail customers don't necessarily plan for the very large utility bills that may result from combined service billing. For example, the City of Philadelphia has a combined water and sewer utility. The staff has found that customers don't differentiate between the water and sewer portions of their bills. The gas works in Philadelphia also is owned by the City, but by a different entity. Proposals have been made to merge the billing of water, sewer, and gas. Customers whose combined water and sewer bill is in the range of $35 to $40 per month might be faced with a peak gas bill of $300 per month during winter months. Such situations create precedence issues. For example, if a customer pays only a portion of a very large combined utility bill, which entity gets to keep and use partial payments?
- **Water is more visible.** Sewer charges and rate increases tend to be less visible than water rate increases. In some cases, this is because portions of the wastewater budget might be grouped with other municipal services, such as fire and police protection. For example, in Alexandria, Virginia, there is a municipal charge included on the property tax bill for maintaining the municipal collection system, in addition to the wastewater charge for treatment and regional collection service that is billed by the Alexandria Sanitation Authority. The full water bill tends to be charged on a single bill to customers; therefore, it is more visible and may seem disproportionately high when compared with some other services that are split among multiple fees, charges, and taxes.
- **New additions to water bills are becoming more common.** Over the past several years, many cities and counties throughout the country have been required by federal stormwater permit requirements to implement more aggressive stormwater management programs. In many cases, this is leading to the development of new stormwater utility fee structures to create a dedicated funding source for stormwater management. While the stormwater charges sometimes are sent out as separate bills, usually it is cost-effective to include these charges on the community's utility bill or property tax bill. In Philadelphia, for example, the $90 million annual costs for the stormwater program are included on the utility bill, along with water and wastewater charges.

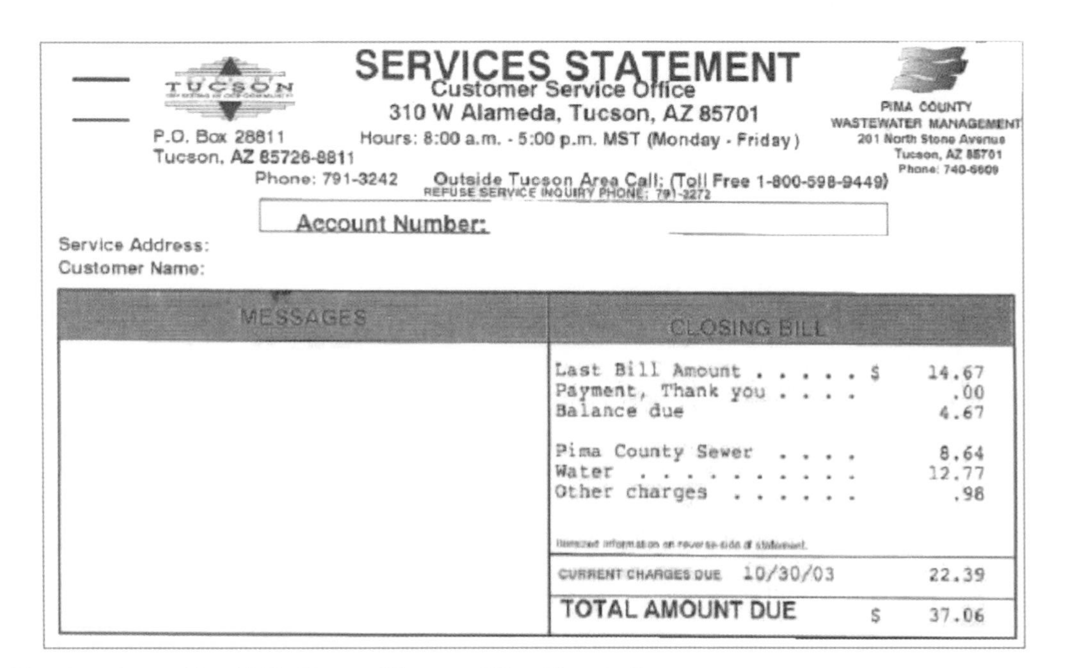

TUCSON

SERVICES STATEMENT
Customer Service Office
310 W Alameda, Tucson, AZ 85701
Hours: 8:00 a.m. - 5:00 p.m. MST (Monday - Friday)

P.O. Box 28811
Tucson, AZ 85726-8811
Phone: 791-3242 Outside Tucson Area Call: (Toll Free 1-800-598-9449)
REFUSE SERVICE INQUIRY PHONE: 791-3272

PIMA COUNTY
WASTEWATER MANAGEMENT
201 North Stone Avenue
Tucson, AZ 85701
Phone: 740-6609

Account Number:

Service Address:
Customer Name:

MESSAGES	CLOSING BILL	
	Last Bill Amount . . . . . $	14.67
	Payment, Thank you	.00
	Balance due	4.67
	Pima County Sewer	8.64
	Water	12.77
	Other charges	.98
	CURRENT CHARGES DUE 10/30/03	22.39
	TOTAL AMOUNT DUE $	37.06

Figure 11 Sample combined bill—Tucson Water and Pima County Sewer

2. Timing of other rate and tax increases

Whether or not billing is grouped, rate increases and tax increases for other utility services or other public services can have a negative effect on a water utility's efforts to secure a rate increase and generate support among customers during its implementation. Cities that have experienced major property tax increases to fund unrelated programs (e.g., a major capital improvement program for schools) will have trouble finding political leaders and customers willing to support even modest water rate increases. When the rate increases do appear on the water bill (e.g., huge rate spikes to implement a mandated combined sewer overflow program), generating support for water rate increases may be even more difficult.

3. Billing frequency

Many water utilities send bills to their residential customers on a quarterly basis. A quarterly bill can be substantial if it also includes charges for sewer, stormwater, and other services. Customers with limited resources who don't plan for these quarterly spikes may find themselves unable to pay a large bill on a quarterly cycle. Though it certainly requires an investment in billing and collections re-programming, many utilities have found it beneficial—both for cash flow and customer budgeting reasons—to switch to monthly billing.

4. Level of detail

Based on input from the research team and utility workgroup that developed this report, there is general consensus that water utility bills benefit from greater detail. The Honolulu Board of Water Supply's experience with a movement toward a more detailed billing format serves as a good example. Figure 12 shows an example of the old bill format, which did not make clear the split between water and sewer charges. Figure 13 shows the new, improved billing format recently introduced. The bills now break out the different charges clearly and give the customer much more information by which they can manage their account. An accompanying "how to read this bill" brochure provides even more information.

Recommendations

Distinguish water charges clearly from other portions of the bill. On combined bills, make sure the water charges are distinguished clearly from other charges, such as sewer and stormwater or wholesale pass-throughs.

Communicate and coordinate with any other utilities for which you provide billing. Build outreach requirements into combined billing agreements. Where multiple charges appear on a single bill, communicate with the staffs and governing boards for the related services about planned water rate increases, and keep up to speed on planned increases in the other charges (e.g., sewerage rate increases planned by a regional wastewater agency whose charges are included on the water bill). Where possible, work together to minimize the cumulative increase to typical customers in a single year. Where substantial increases in a single year cannot be avoided, coordinate communication campaigns so that customers receive the same message from all parties whose charges appear on the bill.

Include enough detail for the average ratepayer to determine how the bill was computed. A typical customer, not schooled in utility rate methodologies and jargon, should be able to compute his or her charge from the information included on the bill.

Consider more frequent billing intervals. If the combined charges of water and other tacked-on services become untenable for customers who don't plan for the uneven quarterly spike of utility charges, consider billing monthly rather than quarterly.

Consider special rate structures and fee options, along with multiple tools and approaches to make sure rates are equitable, defensible, and affordable. The actual rate structure that is used to compute a customer's charge

Water utility bills in many cases could be improved by including more explanatory detail, particularly where there is combined billing.

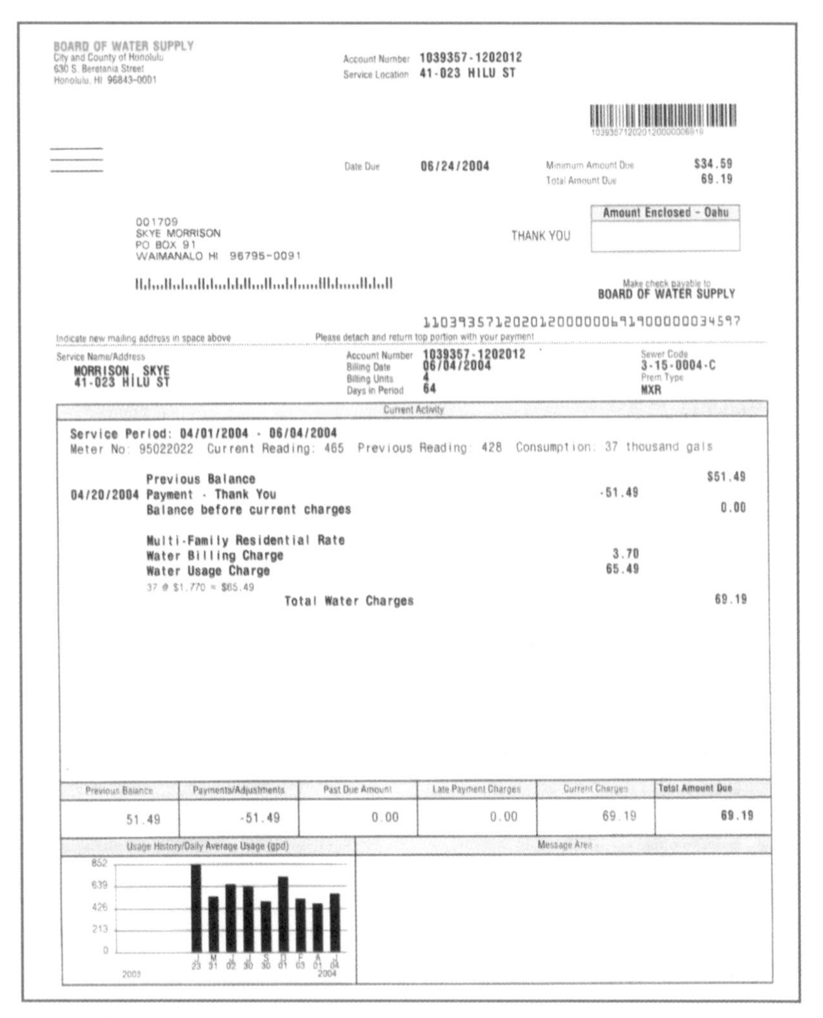

BOARD OF WATER SUPPLY
City and County of Honolulu
630 S. Beretania Street
Honolulu, HI 96843-0001

Account Number 1039357-1202012
Service Location 41-023 HILU ST

Date Due 06/24/2004
Minimum Amount Due $34.59
Total Amount Due 69.19

001709
SKYE MORRISON
PO BOX 91
WAIMANALO HI 96795-0091

THANK YOU

Amount Enclosed - Oahu

Make check payable to
BOARD OF WATER SUPPLY

110393571202012000000691900000034597

Indicate new mailing address in space above

Please detach and return top portion with your payment

Service Name/Address
MORRISON, SKYE
41-023 HILU ST

Account Number 1039357-1202012
Billing Date 06/04/2004
Billing Units 4
Days in Period 64

Sewer Code 3-15-0004-C
Prem Type MXR

Current Activity

Service Period: 04/01/2004 - 06/04/2004
Meter No: 95022022 Current Reading: 465 Previous Reading: 428 Consumption: 37 thousand gals

	Previous Balance		$51.49
04/20/2004	Payment - Thank You	-51.49	
	Balance before current charges		0.00
	Multi-Family Residential Rate		
	Water Billing Charge	3.70	
	Water Usage Charge	65.49	
	37 @ $1.770 = $65.49		
	Total Water Charges		69.19

Previous Balance	Payments/Adjustments	Past Due Amount	Late Payment Charges	Current Charges	Total Amount Due
51.49	-51.49	0.00	0.00	69.19	69.19

Usage History/Daily Average Usage (gpd)

Message Area

Figure 12 Previous billing format—Oahu

can influence the acceptability of rate increases. For example:

- In communities with large low-income or elderly populations, support for rate increases may hinge on developing a **lifeline rate program** or other relief to avoid severe financial hardship for these groups.
- In some communities that have a strong environmental and conservation ethic, or where water is relatively scarce, having a **rate structure that encourages conservation,** such as an increasing block rate structure that charges a higher rate for water purchased above an identified volume, could be helpful in generating customer support for rate increases. In this situation, such conservation rates can demonstrate that the utility is taking action by charging more heavily those customers who are making greater use of a water system, helping to manage demands and reducing unnecessary capital spending.

During the past several years, considerable attention has been devoted in the water industry to identifying and evaluating rate structure options. For example, in 2000, AWWA released an updated edition of its "M1" rate manual,[23]

23. *Principles of Water Rates, Fees, and Charges: AWWA Manual M1*–Fifth Edition, American Water Works Association, 2000.

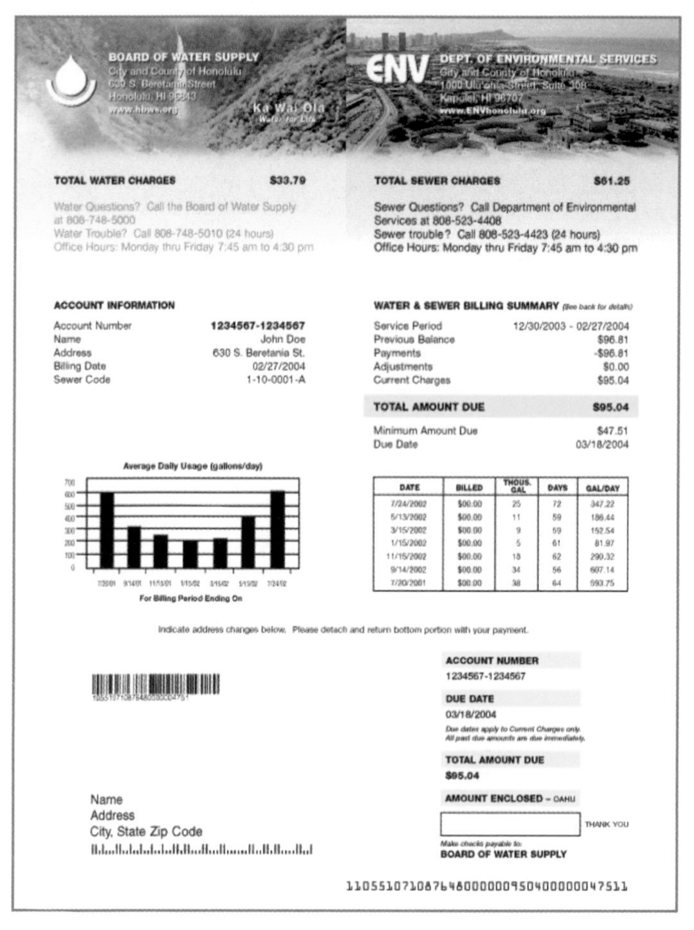

BOARD OF WATER SUPPLY
City and County of Honolulu
630 S. Beretania Street
Honolulu, HI 96843
www.hbws.org

Ka Wai Ola
Water for Life

ENV DEPT. OF ENVIRONMENTAL SERVICES
City and County of Honolulu
Kapolei, HI 96707
www.ENVhonolulu.org

TOTAL WATER CHARGES **$33.79**

Water Questions? Call the Board of Water Supply at 808-748-5000
Water Trouble? Call 808-748-5010 (24 hours)
Office Hours: Monday thru Friday 7:45 am to 4:30 pm

TOTAL SEWER CHARGES **$61.25**

Sewer Questions? Call Department of Environmental Services at 808-523-4408
Sewer trouble? Call 808-523-4423 (24 hours)
Office Hours: Monday thru Friday 7:45 am to 4:30 pm

ACCOUNT INFORMATION

Account Number	**1234567-1234567**
Name	John Doe
Address	630 S. Beretania St.
Billing Date	02/27/2004
Sewer Code	1-10-0001-A

WATER & SEWER BILLING SUMMARY *(See back for details)*

Service Period	12/30/2003 - 02/27/2004
Previous Balance	$96.81
Payments	-$96.81
Adjustments	$0.00
Current Charges	$95.04
TOTAL AMOUNT DUE	**$95.04**
Minimum Amount Due	$47.51
Due Date	03/18/2004

DATE	BILLED	THOUS. GAL	DAYS	GAL/DAY
7/24/2002	$00.00	25	72	347.22
5/13/2002	$00.00	11	59	186.44
3/15/2002	$00.00	9	59	152.54
1/15/2002	$00.00	5	61	81.97
11/15/2002	$00.00	18	62	290.32
9/14/2002	$00.00	34	56	607.14
7/20/2001	$00.00	38	64	593.75

Indicate address changes below. Please detach and return bottom portion with your payment.

ACCOUNT NUMBER
1234567-1234567

DUE DATE
03/18/2004
Due dates apply to Current Charges only. All past due amounts are due immediately.

TOTAL AMOUNT DUE
$95.04

AMOUNT ENCLOSED – OAHU

THANK YOU

Make checks payable to:
BOARD OF WATER SUPPLY

Name
Address
City, State Zip Code

11055107108764800000095040000004751 1

Figure 13 New billing format—Oahu

which devoted more than twenty chapters to descriptions and discussions of various rate structure and fee options. These options range from traditional rate structures, such as uniform rates and declining/inclining block rates to emerging rate options, such as negotiated rates and miscellaneous and special charges.

The current study is not intended to supersede the body of rate study work represented by the update to the M1 manual and related industry documents. Rather, based on the experience of the research team and participating work group utilities in planning and implementing successful rate increases, it is one objective of this report to highlight some of the rate increase options and related technical studies from the options identified in the rate literature that often merit consideration when a utility is faced with implementing a considerable rate increase. The reader is advised to consult additional works, such as the latest edition of the M1 manual, for more details related to these rate structure and study options.

When faced with substantial increases in revenue requirements, water utilities should revisit their rate structures and rate programs. Consideration should be given to the following rate and charge options, which are described in greater detail in Appendix B:

- Special fees and charges (service connection fees, impact fees, and other ways to ensure that the user of those services pays)
- Special dedicated fees or rate components for renewal and replacement
- Increases in the base charge
- Enhanced/revised customer classifications
- Construction works in progress surcharge
- "Lifeline" rates or discounts to protect elderly, low-income or other special groups

Tucson Water's *review of utility costs resulted in the development of a system equity fee.' Capital improvements to provide additional water system capacity have been constructed in advance of when the additional capacity will be fully utilized; current system users have funded these improvements via water rates and have, therefore, provided system capacity to serve future users. By paying the one-time 'system equity fee' when connecting to the water system, a new user 'buys in,' funding his proportionate share of system capacity. The revenues generated by the fee, used to pay debt service, reduce the level of increase in water sale revenue which would otherwise be required.*

- Multi-year rate programs
- "Automatic" rate increases

The potential value of these options depends on factors like the existing rate structure, the customer base, the key stakeholders, and the basis for the increased revenue requirements. However, it is worth at least considering each of these options when faced with major rate increases.

Appendix B also identifies a number of technical studies, such as affordability studies and background information, that can provide the supporting documentation needed to make the case for rate increases. When facing substantial increases, water utilities should consider which, if any, of these technical studies would provide critical supporting information, given the basis for the increase, the decision-making environment, and the stakeholders:

- **Cost-of-service studies** to document revenue requirements and where they should be allocated within the utility's customer base
- **Asset management studies** to document revenues required to maintain the system's assets at a target condition or performance level, and/or cost tradeoffs and efficiencies that result from increased renewal and replacement expenditures
- **Rate impact projections**
- **Multi-year financial models and scenario analyses**
- **Comparisons** with neighboring communities
- **Demonstration of the connection** between sound utility assets and accomplishing other community goals, such as economic development

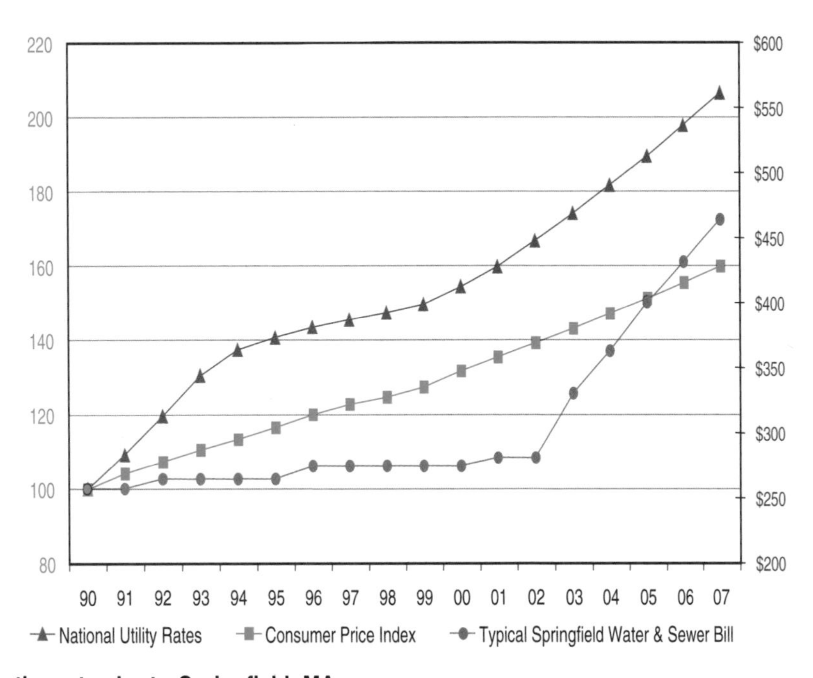

Figure 14 Comparative rate chart—Springfield, MA

- **Affordability analyses** that compare projected charges with household income and other affordability metrics
- **Analysis of historical utility rate increases** vs. increase in Consumer Price Index (CPI), Construction Cost Index (CCI), and other cost indices
- **Pictures or sample sections of tuberculated pipes** or other examples of deteriorating system elements
- **Maps** showing the ages of pipes in various districts or neighborhoods

As one example of how these supporting technical studies can aid in the communication of the need for rate increases, the City of Springfield, Massachusetts, used the analysis of its historical rates in comparison with increases in national utility rates and increases in the CPI and construction costs.

As shown in a figure from Springfield's public outreach brochure (see Figure 14), Springfield's water and sewer rates historically had lagged well behind both national utility rates and the CPI. This was part of the successful rate increase campaign that helped decision-makers and customers understand that Springfield's rates hadn't even kept up with the inflation rate as needed to cover normal increases in salary and equipment costs, much less the costs of renewing and replacing the basic infrastructure that had been neglected for several generations.

"Holding rate levels constant for multiple years does not benefit ratepayers if inflationary increases in operating costs and other expense pressures eventually compound to force a rate increase of such magnitude that ratepayers have extreme difficulty in budgeting for this expense. Such patterns of irregular rate increases increase the risk that ratepayers will pressure rate-makers to resist needed changes, thus increasing credit risk to bondholders."

—James Wiemken, Director, Standard & Poors Credit Market Services

When making such comparisons between a utility's rate history and other measures, it is critical that the rates and other indices being compared be "normalized." Normalization is accomplished either by adjusting all values to an index value at some starting year, or by graphing the dollar figures on a scale that is based on such indexed values, as was done in the case of the Springfield example. That is, in the Springfield example, the scales were set such that the blue line that represents Springfield's rate history and forecast is aligned with both the left-hand index scale and the right-hand dollar value scale.

CONCLUSION

In a recent *Journal AWWA* article, AWWA Executive Director Jack Hoffbuhr noted that, "if water is not valued, it will be wasted without a second thought. For example, the United States, with the least expensive water in the world, has the highest per capita use."[24] As the results of this report have shown, the value of water—or more specifically the public's *undervaluing* of water—figures prominently in any discussion of sustainable ratemaking.

Hoffbuhr continues: "No headway will be made until we find a way to convey the value of water ... A United Nations official said it best: 'Water is the basis of life. Why, then, do we treat water with none of the reverence we have for life?"[25]

The overwhelming weight of evidence shows that utilities with effective, well-planned, well-executed customer communication programs succeed in gaining public support. They defined their objectives, aligned their goals with community priorities, ascertained public attitudes and values, and articulated relevant messages and delivered them effectively.

As the bills come due for meeting increasingly stringent public health regulations and replacing aging infrastructure, utilities that tell their own stories—rather than letting others tell it—will be less frustrated and far more successful.

By focusing on addressing the four key findings of this study—reconnecting people to the value of water, consistently communicating with multiple stakeholders, thinking long-term, and reconciling billing practices and rate structures with community goals—utilities can take control of their own destinies. In making the case for sustainable water rates, utility managers will ensure the proper stewardship of our most critical public resource—our water supplies—and the infrastructure that underpins our civilized society.

24. Hoffbuhr, Jack, "Was Malthus Right?" *Journal AWWA*, August 2003, p. 6.

25. Ibid.

APPENDIX A—CASE STUDIES

As part of the research conducted for this project, the team prepared detailed case studies for a dozen utilities across the U.S. These utilities serve populations ranging from 34,000 to 1.5 million, and the driving forces behind their rate increases have varied from regulatory compliance mandates to the desire for a higher quality of service. The following table summarizes key information from the case studies; full text of the case studies appears on subsequent pages.

Utility	Population served	Drivers for increase	Rate increase	Stakeholders	Analysis tools	Key lesson learned	Communication tools
Alexandria Sanitation Authority, Alexandria, VA	325,000	Consent order compliance mandating a $300 million treatment plant upgrade	84% over first 5 years	City Manager Mayor City Council Bond rating agencies State regulators Customers Chamber of Commerce	Rate study Financing options study Rate impact projections Affordability analysis Rate comparisons	Careful planning and analysis increases the chances of approval— and can help a utility qualify for low-interest loans.	Press releases Mailers Flyers Brochures Board and Council briefings
Augusta, GA	180,000	$300 million capital improvement program to address water infrastructure deterioration, decline in service quality	11% annually, 2000–2007	Staff Mayor County Commission Bond rating agencies Customers	Multi-year financial models Rate comparisons with neighboring communities Maps showing pipe ages in various neighborhoods	To build elected officials' support, distribute planned capital improvements across political jurisdictions—and schedule them to show quick results.	News releases Board meetings Website postings Mailers and fliers Presentations to Commission Public meetings Newsletter
Cleveland Division of Water, Cleveland, OH	1.5 million	$750 million capital improvement program for treatment plant upgrades to address aging, inefficient infrastructure	6.5–7% annually 1996–2000, 3.5% annually 2001–2005	Staff and management Mayor City Council Bond rating agencies City customers Suburban communities	Rate impact projections Multi-year financial models Rate comparisons with similarly sized and neighboring communities Economic impact analysis Affordability analysis Asset management study	Understand, and be extremely sensitive to, the concerns of the ultimate decision-makers.	New releases Press briefings Website postings Fliers Presentations to City Council Visual aids—maps, pictures and samples of tuberculated pipes Field trips

(continued on next page)

Utility	Population served	Drivers for increase	Rate increase	Stakeholders	Analysis tools	Key lesson learned	Communication tools
East Bay Municipal Utility District, Oakland, CA	1.4 million	Cost-of-living salary adjustments; increasing costs for energy and chemicals; rising debt service costs to address increased capital spending for asset renewal and replacement	3.75%	Mayor City Councils Board Bond rating agencies Customers Business community	Rate impact projections Community effects analysis Economic development analysis	Establish a pattern of annual rate increases at or below the rate of inflation.	News releases Editorial board meetings and press briefings Web postings Mailers and fliers Presentations to Council and Board Public meetings and hearings Newsletter
Kenosha, WI	120,000	$29 million new microfiltration plant to replace a filtration plant built in 1917	30% in 1995 (wastewater) 27% in 1999 (water)	Mayor City Council and Board Bond rating agencies State regulators Public service commissions Customers	Rate impact projections Multi-year financial models Rate comparisons with neighboring communities	When public sentiment supports increased spending, be opportunistic—pursue funding for improvements that will support long-term goals.	News releases Presentations to Council and Board Public meetings Special public meetings for senior citizen groups
Las Virgenes Municipal Water District, CA	>85,000	$50–80 million in upgrades required for compliance with federal, state and local regulations, as well as increases in insurance and operating costs	3 steps of 8% over 3 years	Board of Directors Customers	Rate impact projections Multi-year financial models	Provide elected officials clear, relevant information to support rate increases.	News releases Mailers and fliers Public hearings Brochures Newsletter

(continued on next page)

Utility	Population served	Drivers for increase	Rate increase	Stakeholders	Analysis tools	Key lesson learned	Communication tools
Oro Valley, AZ	34,050	$32 million capital improvement plan responding to anticipated potable water demands, as well as compliance with groundwater usage regulations	8.4% in 2004, 4.67% in 2005–2009	Mayor City Council Utility Commission General customers Golf courses, large irrigation users Other industrial/institutional customers Land developers	Rate impact projections Multi-year financial models Rate comparisons with neighboring communities Affordability analyses Analysis of historical utility rate increases versus increases in CIP, CCI and other cost indices	Plan carefully to moderate the size of rate increases, particularly when large capital improvement projects are planned.	Town Hall public meetings News releases Bill stuffers Presentations to Council and Board Public hearings Newsletter
Philadelphia, PA	1.5 million urban, 900,000 suburban	$19.9 million in capital in FY 2002, $26.2 million in FY 2004	5.8% in 2002, 6.8% in 2003, 6.8% in 2004	Mayor City Council Board Multi-family property managers Bond rating agencies General customers Consumer advocacy groups	Multi-year financial model Analysis of historical rate increases versus increases in capital investment programs and other indices	Put ongoing effort into communicating positive stories about what the utility is doing to enhance its performance, control costs, and enhance the community's quality of life.	News releases Web postings Mailers and fliers Fact sheets Public hearings Brochures
Portsmouth, VA	>140,000	Increase in the costs of emergency response and reactive maintenance, combined with an increase in the number of main breaks due to age	32% (water), 76% (wastewater)	Employees Mayor City Council Utility Board Bond rating agencies General consumers Chamber of Commerce Industrial/institutional customers Consumer advocacy groups	Rate impact projections Multi-year financial models Rate comparisons with neighboring communities Asset management study Economic development study Affordability analysis	Demonstrate that the rate increase program will be affordable and is aligned with overall community objectives.	Press briefings Web postings Mailers and fliers Presentations to Council and Board Public information meetings and hearings Brochures

(continued on next page)

Utility	Population served	Drivers for increase	Rate increase	Stakeholders	Analysis tools	Key lesson learned	Communication tools
Springfield, MA	252,000	Federal consent order, aging infrastructure, security needs, and desire for improved efficiency	54% over 5 years	Mayor City Council Business leaders Community activists Wholesale customers	Internal workshop to focus 78th SWSC vision, mission and goals Interviews with external stakeholders to gauge public perceptions of the Commission and its performance, as well as receptivity to the proposed rate increase	Develop messages that speak to your stakeholders' concerns—and communicate them consistently and repeatedly.	Brochure Presentation to Council Public hearing Participation in "Drinking Water Week" booths at local malls
St. Petersburg, FL	240,000	Continuous annual increases in raw water costs, declining consumption due to conservation efforts and a wet weather cycle, low interest rates, badly needed maintenance for the aging infrastructure combine as the drivers	About 15% annually over 5 years (*requested*)	Mayor and City Manager City Council Utility Board Wholesale customers	Rate impact projections Multi-year financial models Rate comparisons with neighboring communities Analysis of historical utility rate increases versus an increase in other cost indices Studies showing the increase in raw water supply costs	Pursue a program of regular annual rate increases. In addition to managing expectations, annual increases can help address uncontrollable changes in operational costs and system revenues.	News releases Mailers and fliers Bill stuffers Presentations to Council Meetings with wholesale customers Public hearings
Tucson, AZ	680,000	Regular rate increases to cover rising costs	4.3%, plus a new system equity fee	Mayor and City Manager City Council Citizens' Water Advisory Committee Customer Rate Design Group	Rate impact projections Multi-year financial models Rate comparisons with neighboring communities Analysis of historical utility rate increases versus an increase in other cost indices	Be thorough in identifying stakeholder groups and developing an understanding of their concerns. Leaving critical stakeholder groups out of the outreach process can delay or derail a rate increase.	Presentations Public meetings Flyers Newsletter

Case Study #1—Alexandria, VA

Name of Utility	**Alexandria Sanitation Authority (ASA), Alexandria, VA**	Median Household Income (annual)	**$67,312**
Population Served	■ **Approximately 325,000 customers served, including 120,000 in Alexandria** ■ **Neighboring Fairfax County, VA owns 60% of the plant's capacity** ■ **Service area of 51 square miles**	Number of Connections	**26,680**
Contact	**Ralph Charlton—Director of Fiscal Services, 703.549.3381, charlton@alexsan.com**	Website	**www.alexsan.com**

Lessons Learned

- Plan carefully, and document the need for—and affordability of—utility rates. This is even more crucial when raising rates after years of no increases.
- Investigate avenues for obtaining low-interest loans, which can lessen the impact of rate increases on customers.

Situation Overview

A mandated $300 million treatment plant upgrade program was the primary influence driving ASA's need for a substantial rate increase. Keeping up with escalating operating costs was a secondary factor. When design costs for the plant upgrade program became available, ASA management made the decision to pursue a rate increase.

The utility's extensive Capital Improvement Program (CIP) largely was due to the implementation of the Potomac Embayment Standards. ASA agreed to comply with these standards; even so, the Commonwealth and the federal government required the Authority to enter into a consent decree.

In 1996, when ASA faced the huge CIP, the utility had not been in the bond market for about 20 years and had not increased rates for 7 years. As a nearly fully developed community, Alexandria had little opportunity to pass along parts of the capital program costs to an expanding customer base.

Actions Taken

To plan for compliance with the $300 million upgrade program, ASA commissioned a rate study that was designed to explore a variety of policy issues, rate structures, and financial planning options. For example, the study fully explored both the municipal bond market and state revolving fund (SRF) loan programs as vehicles for funding the capital program. The study also investigated financing options related to Fairfax County, which by contract has a 60 percent stake in ASA's capital facilities. The presentation graphic below illustrates the range of options considered for financing alternatives that included Fairfax County's share of the capital requirements in ASA's bonding program.

The required rate increase was expected to be significant. To build support for the increase, ASA analyzed the results of numerous research efforts. The utility also conducted rate impact projections, ran multi-year financial models, compared economic data with neighboring communities, and conducted affordability analyses comparing projected charges with household income and other affordability metrics.

After fully exploring rate and financing options, ASA realized that rates would need to nearly double during the first 5 years of its rate increase program. A rate study report recommended a cumulative rate increase of approximately 85 percent over the first 5 years.

When technical studies were completed, ASA began an aggressive campaign to brief key stakeholders in the decision-making process. ASA identified the following groups as stakeholders in the approval of the rate increase:

- City Manager
- Mayor
- City Council
- Bond rating agencies
- State regulators
- General customers
- Chamber of Commerce

ASA communicated to target audiences and stakeholder groups using press releases, mailers, fliers, and brochures. In addition to briefing its governing board in a series of meetings, ASA's staff and rate team held a special briefing of the Alexandria City Council, so that Council members were aware of the timing and rate impacts of the mandated program.

Rate Structure Implemented

The rate increase that ASA's staff and rate team recommended was adopted by the Board of Directors. The implemented rate increase program raised rates 84 percent over the first 5 years, including an initial rate increase of 30 percent, followed 6 months later by a second rate increase of 28 percent. The rate of $1.84/1,000 gallons increased to $3.06/1,000 gallons. By the end of the 5-year period, the typical residential bill (wastewater only) rose from $32.72 per quarter to $59.62 per quarter.

The initial set of rate increases has been followed by a series of lesser increases required to complete program construction, but also to address normal increases in operating costs.

Conclusion

ASA did not encounter organized opposition to the rate increase, and election cycles were not a factor. Careful planning, which documented the need for—and affordability of—the program, helped ASA qualify for substantial amounts of low-interest loans, which have lessened the program's impact on ASA's customers. To date, the program has qualified for more than $99 million in low-interest SRF loans, and the advance rate and financial planning have been key elements in securing this funding.

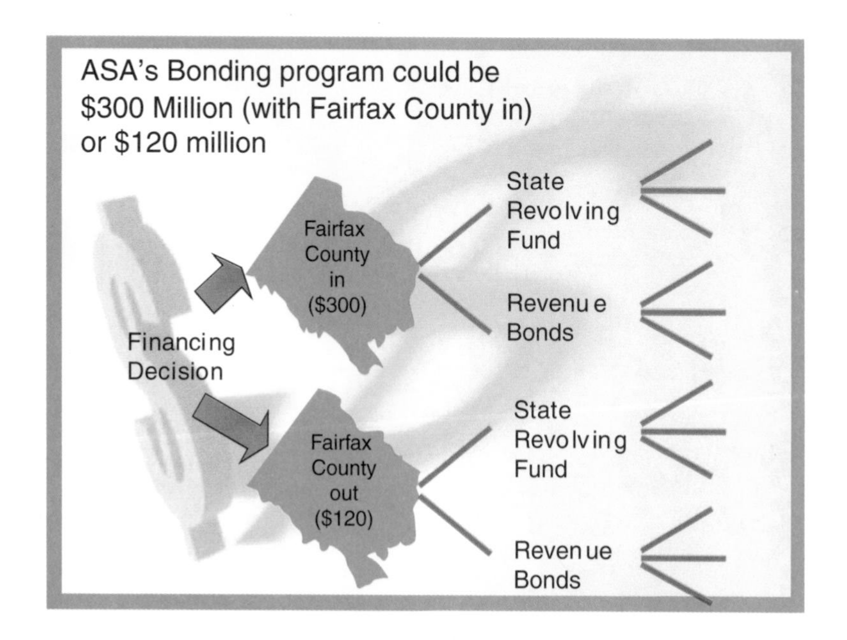

Graphic from presentation prepared to support ASA's requested rate increase

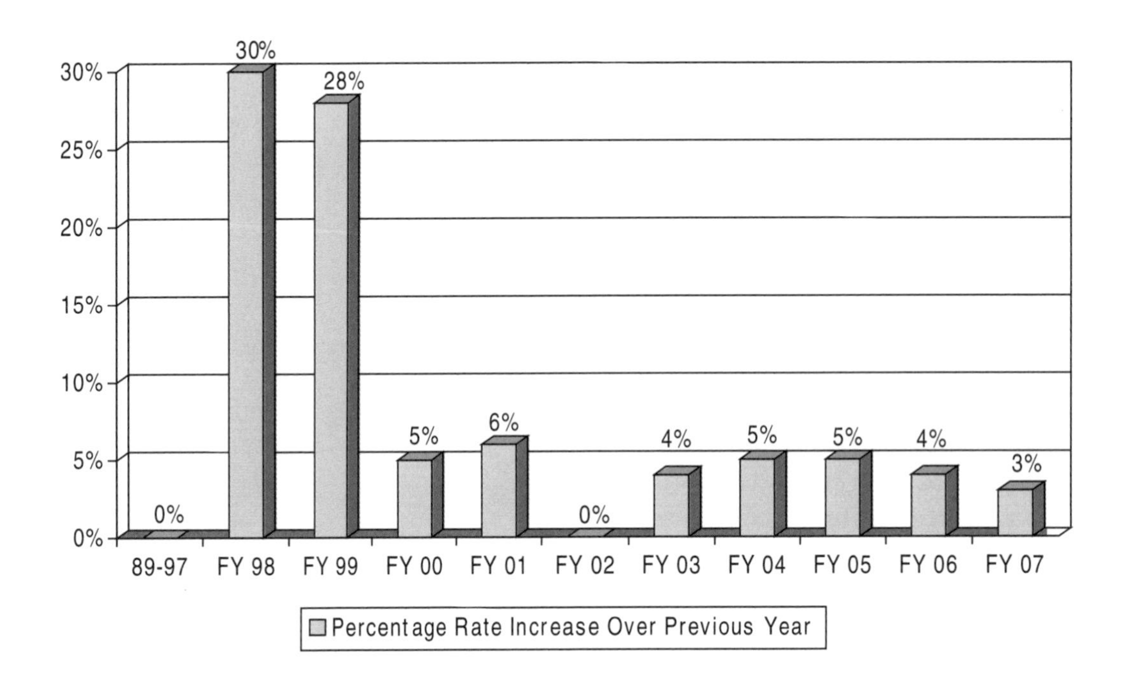

Past and planned rate increases at ASA

The Alexandria case is unique in that the utility was driven to modernize its treatment facility by the Commonwealth and federal government mandates. In this region, the care and state of the Potomac River and the Chesapeake Bay are highly visible, potentially controversial political topics. Making improvements to a sewage treatment facility that empties into to a tributary of the Potomac River would be a difficult effort to oppose successfully in this environment.

Case Study #2—Augusta, GA

Name of Utility	Augusta Utilities Department, Augusta, GA	Median Household Income (annual)	$30,399
Population Served	180,000	Number of Connections	67,000
Contact	Max Hicks—Utility Director, 706.312.4160, mhicks@augustaga.gov	Website	www.augustaga.gov/departments/utilities

Lessons Learned

- Make a compelling argument for rate increases in clear terms that your stakeholders can understand.
- To build elected officials' support, distribute planned capital improvements across political jurisdictions to show quick results.
- Develop a long-term financial plan to manage multi-year rate increase requirements.
- Work to establish legal protections against large transfers of utility cash to general funds.

Situation Analysis

The City of Augusta and Richmond County formed a consolidated government (Augusta, GA) in 1996. Existing water and wastewater utilities for the City and County were consolidated as well, and the utility was reorganized to serve the needs of the combined government's constituency. In both systems, repair/rehabilitation and maintenance historically had been significantly underfunded, raising the risk of system failures. Soon after the consolidated utility was put in place, a severe drought and unprecedented system demands overtaxed the system, contributing to acute operating issues, numerous line breaks, and regular service interruptions.

Questioned by the Augusta-Richmond Board of Commissioners on the system problems, Utility Director Max Hicks responded that although rate collections were on target, there were inadequate funds available to maintain the system properly and fund needed capital projects—primarily due to large transfers of utility operating revenues to the general fund over the previous several years.

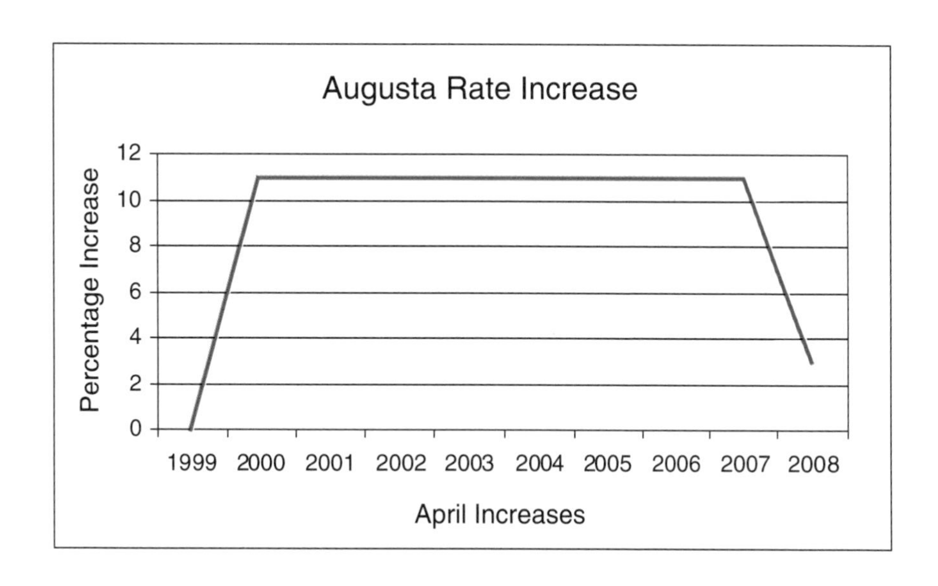

During its May 1998 term, the Richmond County Grand Jury formed a subcommittee to examine the system failures and their causes, and to make recommendations for improvements to prevent future occurrences. One of the Grand Jury's recommendations was that a Finance Officer be appointed for the Utilities Department. The utility was authorized to hire a finance director, who was then a leader in the investigation of historic spending and fund transfers from the City and County utilities' pooled cash fund.

An exhaustive investigation identified transfers totaling $44 million from the City water works fund during the 8 years preceding government consolidation. The bulk of these transfers had been made during a mayoral administration characterized by a pledge not to raise taxes. Utility funds were subsequently reallocated to riverfront development and other community amenities. An additional $30 million was transferred from the water works fund as "payment in lieu of taxes."

Actions Taken

As the grand jury investigation was under way, the utility continued to educate the Commissioners on the extent of under-funding of system maintenance and repair. Concurrently, the utility began an aggressive program of master and capital planning to define what was needed to repair and rehabilitate the system. Projected operations and maintenance costs also were analyzed to define realistic annual O&M budget needs. Planning activities culminated in a $280 million Capital Improvements Plan. Rate analyses were performed to assess affordability and explore different rate increase options.

Logo for Augusta Utilities Department's Benchmark 2010 improvements program

In addition to planned bond funds, a series of significant annual rate increases was required to fund the CIP: 11 percent annual rate increases through 2007, with annual increases of 3 percent from 2008 through 2010.

To build support for these increases, the utility worked closely with the Board of Commissioners and educated them on system needs. Central to the utility's communication strategy was the identification of high-priority projects in each political jurisdiction; these projects were included in the CIP so that elected officials could demonstrate results quickly after the rate increases had been put place. The support from the Commissioners has been a vital element in the utility's progress.

Rate Increases Approved

The utility retained an underwriter and issued bonds to fund system upgrades. The first major bond issue was in 2000 ($90 million), followed by a second in July 2002 ($148 million). In concert with the first bond issue, the City put a program of rate increases in place. Rates were increased 11 percent in April 2000, and the rates have increased at a fixed rate of 11 percent each year since. The annual 11 percent increases are planned through 2007, with annual rate increases of 3 percent from 2008 through 2010.

Conclusions

Even with the rate increases, Augusta Utilities Department still has rates that are among the lowest in the state—in the bottom third, even with increase projections.

Subsequent to the rate increases, the Department has leveraged its capital program into an extended opportunity to broadcast positive news on the utility's forward progress. A brand image and logo were developed to facilitate communication regarding the utility's overall improvements program, dubbed Benchmark 2010. The Benchmark 2010 program and its separate components have been publicized on the utility's website, in numerous press releases and flyers, in presentations to community groups, and on the local cable television access channel. On-site signage alerts residents to Benchmark 2010 project activities and progress.

Major project activities such as an extensive Automated Meter Reading (AMR) implementation program have received positive media coverage, and Director Hicks and the utility continue to receive unprecedented support from the community. Benchmark 2010 embodies the Department's goal to become a preferred water and sewer service provider—and a utility from which other systems will want to learn because of its demonstrated value to the community.

"Everyone realizes that our motto, 'Water Is Life,' is indeed a truism," said Utility Director Max Hicks. "Dependable water service at an economical cost is vital to the life of a community."

Case Study #3—Cleveland Division of Water

Name of Utility	City of Cleveland Division of Water, Cleveland, OH	Median Household Income (annual)	$36,000
Population Served	1.5 million	Number of Connections	414,000
Contact	Julius Ciaccia—Commissioner of Water, 216.664.2444, julius_ciaccia@ClevelandWater.com	Website	www.clevelandwater.com

Lessons Learned

- Understand, and be extremely sensitive to, the concerns of the ultimate decision-makers (in Cleveland's case, the Mayor and City Council).
- Plan carefully to distribute the costs of major capital programs over several years, lessening the annual impact on customers.
- Use visual aids (such as sections of tuberculated pipe) and facility tours to raise awareness of system conditions.
- Maintain open, honest relationships with the media and community decision-makers to build mutual trust.

Situation Overview

Water for the City of Cleveland is drawn from Lake Erie through a system of 8- to 10-foot intake tunnels, through treatment facilities, pumping stations, and finally into a distribution network that delivers water to over 400,000 customers. The entire distribution system is more than 5,000 miles long (water mains alone); end–to-end, this system would stretch from Cleveland to San Francisco and back again.

In the late 1990s, the City of Cleveland Division of Water (CWD) needed to launch a major capital improvement program to renovate its four water treatment plants. This endeavor, named the Plant Enhancement Program (PEP), is a 10-year, $750 million project, the largest investment since original plant construction. Major facility planning for this project was completed in 1999 with projects prioritized into three phases. Currently the Division is completing construction on Phase I projects and is in the design stage for Phase II projects.

Like many utilities, the Division of Water historically has faced political resistance to water rate increases. Although the Division's service area has grown outside the City proper and now encompasses 640 square miles in 4 counties serving 72 suburban communities directly, and 6 on a wholesale basis, the approval of the Cleveland City Council is needed for any change in water rates.

Actions Taken

Primary stakeholders for the rate increases were identified as:

- CWD employees and management
- Mayor
- Cleveland City Council
- Bond rating agencies
- City customers
- Suburban communities

The Division of Water used a variety of analyses and research to support the increase effort. This included rate impact projections, multi-year financial models, rate comparisons with similar sized and neighboring communities, economic impact, affordability metrics and asset management studies.

To communicate the supporting evidence to stakeholders and decision-makers, the Division of Water employed a range of communication tools, including new releases, press editorials, press briefings, website postings, fliers, and presentations to City Council. In addition, the Division of Water used visuals to demonstrate the condition of the water system. The visuals included maps, pictures and samples of tuberculated pipes as well as field trips for Council members to see first-hand the condition of its treatment plants.

CWD also has received national media attention for its water system from two major events. In January 2000, the Division experienced a major water main break in its downtown area on a 75-year-old pipe, causing flooding of city streets and disruption of water service. The Division also made headlines during the August 2003 power blackout, when all four of its treatment plants lost power and drinking water service was curtailed to most of its service area.

Rate Structure Implemented

Since 1991, the Division of Water has tried to implement multi-year rate increases to avoid yearly sessions at City Hall. During the period from 1993 to 1996 rates were passed that established 8.5 percent annual increases. From 1996 to 2000, water rates increased on the average 6.5 percent to 7 percent annually. Currently, the Division is in the middle of its latest rate period (2001–2005) that calls for 3.5 percent annual rate increases.

As can be seen from the graph below, CWD's multi-year rate increases have decreased, such that now they are basically keeping up with inflation. Currently, CWD is planning for its next water rate increases for the planning period 2006–2010.

The Division of Water implemented special rate provisions for small consumers, the elderly, and low-income permanently disabled individuals. For small consumers, the Division priced the first million cubic feet (mcf) at a substantially lower rate than subsequent mcfs. For the elderly and low-income permanently disabled, the Division offers a Homestead Exemption rate that provides reduced rates to homeowners who meet annual income and age limitations.

Conclusion

Although the identified groups and stakeholders were important for a successful rate increase implementation, the most critical stakeholders were the ultimate decision-makers: the Mayor and City Council. These were the stakeholders given the most focused attention. In addition, because initiating rate increases is a major political issue in Cleveland, election cycles were a factor considered in planning for the rate increases.

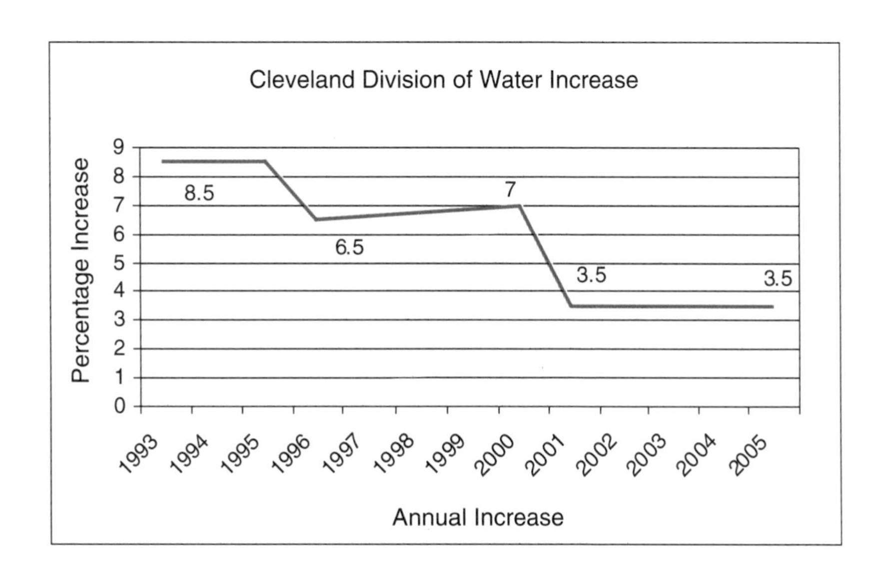

Two major events—a large downtown water main break in 2000 (left) and the August 2003 power blackout (above)—have raised greater awareness of the need for continued investments in the City's water system.

There was no organized opposition to the Division of Water's rate increase, although several suburban mayors voiced their opinions at City Council meetings. The rate increase was justified as needed primarily to raise capital for continuing the Plant Enhancement Program.

The Division of Water builds community support through continuous communication with the media and the public. The utility's website contains extensive historical information on the utility system, as well as detailed current information for consumers.

Case Study #4—Oakland, CA

Name of Utility	East Bay Municipal Utility District, Oakland, CA	Median Household Income (annual)	■ Contra Costa County: $45,087 ■ Alameda County: $37,544
Population Served	■ 1.4 million ■ Service area of 325 square miles	Number of Connections	375,000
Contact	Gary Breaux—Director of Finance, 510.287.0310, gbreaux@ebmud.com	Website	www.ebmud.com

Lessons Learned

- Establish a pattern of annual rate increases that are at or below the rate of inflation.
- Work diligently to understand stakeholders' priorities and concerns. Focus groups and polling have proven to be effective tools.
- Use many different communication tools and avenues to inform the public about the utility's activities, issues, and successes.
- Keep the utility's website information up-to-date, and use it as a means of publicizing rate schedules and progress on capital projects.

Situation Overview

The California legislature formed the publicly owned East Bay Municipal Utility District (EBMUD) under the Municipal Utility District Act in 1921. Today, the utility supplies water and provides wastewater treatment for the San Francisco area, including parts of Alameda and Contra Costa counties in northern California.

A seven-member board approves EBMUD's policies; wards elect the board members to 4-year terms. The policies are implemented under the direction of the General Manager. Each board member sits on a standing committee that makes recommendations on policies and improvements for the entire board's consideration. Board meetings are open to the public, and minutes are published on EBMUD's website.

Revenues come from a variety of sources, including sales of water and hydroelectric power, and customer rates. Customers pay a number of different charges including meter service charges, sewage treatment charges, and a wet-weather facilities charge. In addition to service charges, EBMUD receives funding from property taxes.

For more than a decade, the utility has made a practice of implementing modest annual rate increases that are at or below the rate of inflation.

In 2003, the subject year for this case study, the utility needed to implement a rate increase to address:

- cost-of-living salary adjustments for utility employees;
- increases in the cost of energy and supplies such as chemicals; and
- increasing debt service costs to:
  - —support the utility's capital program to maintain its facilities in good repair, and
  - —develop supplemental water supplies to lessen rationing during times of drought.

Actions Taken

Because of its practice of annual increases, the utility was able to construct a rate plan that called for a moderate increase to help respond to rising costs. The increase averaged about 3.75 percent across the board for all customers.

EBMUD identified the following key stakeholders for outreach related to the rate increase:

- Mayors of cities served
- City Councils of cities served
- Utility Board
- Bond rating agencies
- Customers
- Business community

Utility information piece produced annually by EBMUD

For public education to support the rate increase, EBMUD conducted a variety of studies and analyzed available research. Analyses included rate impact projections, as well as research on community effects and economic development.

EBMUD relied on a number of different tools and approaches to communicate with the stakeholders and the general public, including news releases, editorial board meetings, press briefings, detailed website postings, mailers, Q&A sheets, presentations to the City Councils and the Board, public meetings, public hearings, and newsletters. The utility also produces a high-quality publication titled "All About EBMUD," updated annually, with detailed information on the water and wastewater systems, water consumption and conservation, and historical system information.

The utility uses its website as a primary source of information, with detailed information on capital project implementation and minutes from public meetings. Detailed schedules for rates, charges, and fees also are published on the website, and site information is updated continuously.

Among EBMUD's stakeholders is a consumer watchdog group called WATER, which describes itself as an independent monitor of the quality of service provided by EBMUD and other utilities. In contrast to the group's criticism of other entities it monitors, WATER representatives have been complimentary of EBMUD's concerted efforts at community outreach. "The utility has generally been good to work with and has kept us informed," said WATER representative Bill Highfield in interviews for this case study.

Rate Structure Implemented

A rate increase of 3.75 percent for all customers was approved in June 2003. The average annual bill for residential customers is $324, based on 100 cubic feet of usage; the average household uses 1,100 cubic feet per month. For the average residential customer, this rate increase will raise monthly water rates only slightly—from $26.09 to $27.00.

Conclusion

Through the years, EBMUD has built significant goodwill in the community by continually engaging its stakeholders, monitoring and responding to public opinion, and consistently raising rates by a moderate amount each year. The annual rate increases have enabled the utility to lessen the rate impacts of major capital programs, such as a $200 million, 10-year program of seismic upgrades in the 1990s.

“We’ve never had problems in terms of overt opposition to rate increases—even the one to fund seismic upgrades following the Loma Prieta earthquake,” said Utility Finance Director Gary Breaux. “We’ve tried every year to implement increases at or below the rate of inflation, and to communicate what we were doing. Even before websites, we used public meetings and outreach to potential opponents, City Councils—anybody who had an interest or a stake in what the utility does.”

Case Study #5—Kenosha, WI

Name of Utility	Kenosh Water Utility, Kenosha, WI	Median Household Income (annual)	$46,970
Population Served	■ 120,000 (grown 10,000 since 1999) ■ Service area of approximately 100 square miles	Number of Connections	28,500
Contact	Ed St. Peter—General Manager, 262.653.4305, ed.st.peter@kenoshawater.org	Website	www.kenosha.org

Lessons Learned

- When current events increase the awareness of system vulnerabilities, be opportunistic—pursue funding for improvements that will position the utility for long-term safety, security, compliance, and cost-effectiveness.
- Make a showcase of major system improvements to increase community confidence and trust.

Situation Overview

Kenosha Water Utility (KWU) is a municipally owned, fiscally independent public utility, organized by authority of the Wisconsin Legislature. A board comprising six City-appointed alderpersons governs KWU, and utility revenues are derived solely from service charges.

KWU provides water and wastewater services to a growing population of more than 100,000, with a service area that includes the City of Kenosha, the Village of Pleasant Prairie, the Town of Somers, and the Town of Bristol. The entire water supply comes from Lake Michigan via two intake valves. The water distribution system is extensive, consisting of 326 miles of water main, 5,123 line valves, 2,919 fire hydrants and more than 28,000 active service connections.

The chief driver for the subject rate increase was the need to upgrade the utility's only water production plant—a filtration plant constructed in 1917. In the wake of the 1993 *Cryptosporidium* outbreak in nearby Milwaukee, which sickened thousands of consumers, KWU was particularly interested in expediting system improvements to reduce the risks of a similar occurrence.

Actions Taken

To replace its aging filtration plant, KWU opted to construct a new microfiltration plant with a capacity of 21.7 million gallons per day (MGD), at an estimated cost of $29.5 million. The planned facility would provide treatment levels well beyond what existing regulations required, but the utility felt the extra assurance was important to safeguard its customers.

To fund the plant, the utility developed an implementation program that called for rate increases in two phases—first in 1995, and again in 1999. These two increases addressed wastewater and water rates, respectively.

KWU identified several key groups and stakeholders as targets for its public outreach efforts:

- Kenosha Mayor
- City Councils
- Utility Board
- Bond rating agencies
- State regulators
- Public service commissions
- Industrial and institutional customers

To support the outreach efforts, KWU conducted analyses and studies including rate impact projections, multi-year financial models, and rate comparisons with neighboring communities.

KWU communicated with stakeholders via news releases, presentations to the City Councils and the Utility Board, and a series of public meetings—including special public meetings to hear concerns from senior citizen groups.

Rate Structure Implemented

KWU implemented the designed program of two rate increases in two separate years. The first increase in 1995 of 30 percent was for wastewater. The second increase in 1999 of 27 percent was for water. The utility took election cycles into account, and rate increases were planned to avoid Board member election years.

In June 1999, the O. Fred Nelson Water Production Plant opened as the largest operating microfiltration drinking water plant in North America. Within 5 days, more than 700 people visited the new plant for self-guided tours, featuring 17 different information stations.

O. Fred Nelson Water Production Plant

Conclusion

Unlike previous rate increases in 1992, 1994, and 1995, this increase was presented as a unique event tied to the Milwaukee 1993 *Cryptosporidiosis* event that sickened thousands of consumers. This increase was intended strictly to raise the capital to build the new microfiltration plant. Considering the recent publicity around the Milwaukee *Cryptosporidium* outbreak, along with KWU's outreach efforts to communicate system needs, the rate increase program encountered no organized opposition. The utility subsequently has continued to publicize the success of its innovative microfiltration plant; to date, thousands of visitors have toured the facility.

Case Study #6—Las Virgenes, CA

Name of Utility	Las Virgenes Municipal Water District, Calabasas, CA	Median Household Income (annual)	$185,000+
Population Served	Over 85,000	Number of Connections	19,500
Contact	Arlene Post, Director of Resource Conservation and Public Outreach, 818.251.2100, apost@lvmwd.com	Website	www.lvmwd.com

Lessons Learned

- Know your audience, and understand what is important to them. Rate increases can be controversial no matter what the demographics of your customer base.
- Conduct the analyses necessary to support required rate increases. Summarize the findings to provide elected officials accurate and relevant information to share with constituents.

Situation Overview

The Las Virgenes Municipal Water District serves residents in communities that are part of western Los Angeles County, including Agoura Hills, Calabasas, Hidden Hills, and Westlake Village. The utility provides potable water, recycled water, wastewater, and biosolids composting services.

A five-member, publicly elected Board of Directors governs the District. Each member represents one of five geographic areas, and the members hold overlapping, 4-year terms. The Board appoints a General Manager, who oversees utility operation, and an attorney.

The majority of the District's customers enjoy lifestyles that can be described broadly as upscale; many of the homes, for example, are located on multi-acre sites, with an average cost of more than $800,000. While the cost of water is not a major factor in the day-to-day lives of most District customers, raising utility rates is a politically charged issue.

In recent years, the District has faced a growing funding gap in segments of its budget, resulting from:

- capital improvements to comply with escalating regulations, both in water and wastewater;
- escalating costs for insurance, electricity, and chemicals;
- enhancements to system security; and
- updates to aging infrastructure and to strengthen facilities' ability to withstand seismic events.

Capital costs over a 5-year horizon were estimated at nearly $60 million and could run higher under specific regulatory scenarios.

Actions Taken

Las Virgenes began planning over a year prior to projections for a rate increase in fall 2003. The utility estimated that impacts per household could reach as high as a 60 percent increase—a sum the Board and staff viewed as significant, even considering the demographics of District customers. Several Board members had run on "Low Rate" platforms.

Given these dynamics, the staff engaged the Board early and actively in rate discussions. Extensive data and background was provided on reasons behind higher costs, efforts to mitigate cost impacts, and options for rate changes. A key message to Board members was desirability of greater control of the utility's destiny in the inevitable environment of greater regulatory requirements and added costs.

To mitigate the impact on customers, the District proposed a series of staged increases over a 3-year period. In addition, customers with exceptionally low water use (12 units or less in a billing period, a single unit being equal to 748 gallons) would receive a 10 percent discount on their sewer charges.

To help Board members explain the rate change to constituents, utility staff provided the Board detailed analyses, including rate impact projections, multi-year financial models, and impacts of uncontrollable costs like regulatory compliance and utilities. The District also prepared information to facilitate Board members' responses to constituents' questions and concerns about the rate increases.

The utility conducted significant outreach to communicate with customers about the nature and need for the rate change, using tools such as news releases, public hearings, brochures, and newsletters. Special efforts went to inform employees, who serve as a conduit into the community. Key messages to customers stressed the District's cost-cutting efforts and that fees for environmentally sound sewer service would remain below $1 per household, per day.

Rate Structure Implemented

The Las Virgenes Board approved the staff-recommended increase. No organized residential opposition was mounted, and only a few dozen calls were received—some complaints, and many seeking additional details about the rate change.

Conclusion

Knowing your audience is crucial. In Las Virgenes's case, the focus was to fully inform customers through varied outreach methods and to make certain that the five Board members were fully involved in forming the rate change, then had information and materials to assist them in communicating with constituents about the new rates and the reasoning behind them.

EVERY day, 24/7, 365 days a year, Las Virgenes Municipal Water District is treating wastewater from more than 85,000 local residents. Our service is transparent, safe, environmentally friendly and energy efficient — and that's the way it should be.

Costs to provide service, however, have nearly doubled in the last 10 years—with skyrocketing energy, environmental, security and insurance prices. Cost containment efforts are no longer enough, making it necessary to raise customer charges.

WHAT ARE THE CHANGES?

Sewer fees will continue to be billed every other month, as part of your water bill. To keep fees as low as possible, the increases will be stepped in over time. On January 1, 2004, sewer charges will go up $2 per month; then $1 per month on July 1, 2004; and $1.50 per month, a full year later, in July 2005. Customers with exceptionally low water use (12 Units* or less in a billing period) will receive a 10% discount on their sewer charges. (*A Unit is a measure of water use equal to 748 gallons.)

It will still cost less than $1 per day to have all the wastewater from your household treated and disposed, even after the changes take effect.

FAST FACTS ABOUT SEWER SERVICE

- 15,958 sewer connections in the LVMWD service area
- 56 miles of trunk sewer lines operating 24/7
- 9 million gallons of wastewater safely treated each day
- 75% (or almost 7 million gallons per day) of treated wastewater recycled as irrigation water
- ZERO gallons of recycled water discharged into Malibu Creek April 15 to November 15
- 6,650 wet tons of solid waste removed from wastewater each month and beneficially composted
- Less than $1 per day for a household's sewer service

SEWER SERVICE CHARGES

How your sewer service rates will be changing and why.

Las Virgenes Municipal Water District
4232 Las Virgenes Road, Calabasas, CA
(818) 251-2200 www.LVMWD.com

10/03 VerU1 Mdp

Las Vigenes Municipal Water District brochure explaining new sewer service charges

Case Study #7—Oro Valley, AZ

Name of Utility	**Oro Valley Water Utility, Oro Valley, AZ**	Median Household Income (annual)	**$61,037**
Population Served	**34,050**	Number of Connections	**16,133**
Contact	**Shirley Seng—Utility Administrator, 520.229.5013**	Website	**www.ci.oro-valley.az.us/ WaterUtility**

Lessons Learned

- Plan carefully to moderate the size of rate increases and avoid customer rate shock, particularly when large capital improvement projects are planned.
- Use bond funds, when possible, to help reduce the size of rate increases and spread increases over a longer period of time.

Situation Overview

The Town of Oro Valley is a fast-growing community adjacent to Tucson, Arizona. Historically, the Town has depended solely on groundwater for its water supply, and turf users have accounted for about 30 percent of potable water demands.

The need for increases in water utility rates were driven by two interdependent influences:

- the expectation that ongoing growth would continue to increase demand on the water supply and system; and
- the need to remain in compliance with Arizona Department of Water Resources regulations relating to groundwater levels and allowable groundwater usage.

In response to these pressures, the Water Utility Commission mandated capital improvements to its potable water system—and the construction of a new reclaimed water system to serve irrigation needs.

In addition, the Commission opted to impose a dedicated fee to fund a portion of the reclaimed water system development costs. Initially, this fee raised stakeholder concerns about equity of the distribution of reclaimed water system costs between potential reclaimed water users, potable water customers, and developers.

Actions Taken

The utility's two-phased Capital Improvements Plan (CIP) called for $32 million in funding over a 5-year timeframe. To fund Phase One of its CIP, the utility requested a rate increase of 8.4 percent (including a new dedicated fee for the reclaimed water system) for fiscal year 2004. This initial increase would be followed by annual rate increases of 4.7 percent from FY 2005 through FY 2009; the current financial plan does not anticipate further rate increases after FY 2009.

To support the rate increase, the utility conducted a variety of studies and analyses: potable water system and reclaimed water system master plan CIP requirements; rate impact projections; multi-year financial models; rate comparisons with neighboring communities; affordability analyses that compared projected charges with household income and other affordability metrics; and analysis of historical utility rate increases versus increases in other cost indices.

In conducting outreach to build support for the needed increases, the utility targeted the following groups:

- Mayor
- City Council
- Utility Commission
- General customers
- Golf courses, large irrigation establishments and other industrial/institutional customers
- Land developers

The utility used several methods to support its communication program, including Town Hall public meetings, news releases, bill stuffers, presentations to the City Council and utility board, public hearings, and newsletters.

Rate Structure Implemented

The initial rate increase for Phase One of the CIP was approved, resulting in a rate increase of 8.4 percent in fiscal year 2004. This increase includes the implementation of a new Groundwater Preservation Fee (GPF) of $0.21 per thousand gallons, which will be collected from all water customers—both potable water and reclaimed water users—to help pay for implementation of the new reclaimed water system. In addition, a reclaimed water volume charge was implemented. Future rate increases will be subject to review by the Utility Commission and the Town Council on an annual basis.

The utility rates do not include additional dedicated fees for rehabilitation and repair, nor do they include lifeline rates or discounts for the elderly or other special groups. No changes were made to the level or structure of the Town's water system development fees, which already included a small component for renewable water projects like the new reclaimed water system. In planning future water rate increases, the utility intends to remain highly competitive with rates in neighboring communities.

Conclusion

The Oro Valley Town Council and the Water Utility Commission were the ultimate decision makers for the rate increase. Utility staff dedicated themselves to the completion of lengthy processes to plan, educate the public, and obtain approval for the rate increase. There was no notable opposition to the rate increase or additional fee. Election cycles were considered in planning for the increase; increases in other utility rates were not considered explicitly.

The Oro Valley Water Utility's approach to rate-setting reflects careful planning to avoid customer rate shocks, despite large capital improvement projects. "Oro Valley reviews and evaluates its water rates annually using 5- and 10-year projections," said Utility Administrator Shirley Seng. "Small rate increases have been implemented annually since 1996, when the Town acquired the utility. The majority of our capital improvements are funded with bonds financed over the life of the facilities. This allows the utility to spread the capital costs over time and over a larger customer base."

Case Study #8—Philadelphia, PA

Name of Utility	City of Philadelphia Water Department, Philadelphia, PA	Median Household Income	$30,746 (urban)
Population Served	■ 1.5 million (Urban) ■ 900,000 (Suburban)—200,000 water and 700,000 wastewater	Number of Connections	473,000
Contact	Bernard Brunwasser, Deputy Water Commissioner—Finance and Administration, 215.685.6106, bernie.brunwasser@phila.gov	Website	www.phila.gov/water

Lessons Learned

- Spread increases over multiple years to make annual impacts "bite-sized."
- Put ongoing effort into communicating positive stories about what the utility is doing to enhance its performance, control costs, and enhance the community's quality of life. Rate increases are easier to accept when the public is aware of the value the utility is providing.
- To maintain realistic expectations, budget conservatively—and don't shoot for the absolute minimum in rate increases.

Situation Overview

The City of Philadelphia Water Department (PWD) provides water, sewer, and stormwater services to a sizeable urban and suburban population. The utility has aggressively pursued greater efficiency over the past decade via measures such as automated meter reading, improved leak detection, and regular management audits. Yet PWD also operates under some of the industry's most challenging administrative constraints, including an adversarial formal review process for all rate increase requests.

By fiscal year 2001, rates had not been increased for 6 years—and utility was experiencing a combination of circumstances that made a rate increase crucial:

- Increased costs for compliance with regulations such as the Enhanced Surface Water Treatment Rule and Combined Sewer Overflow requirements
- Large increases in labor costs due to union contract obligations—a 27 percent increase in salaries and 45.3 percent increase in fringe benefits since 1992
- A major increase (23.1 percent) in utility operating costs and materials
- A 16.9 percent increase in mandated annual inter-fund payments to the City's General Fund
- An increase in capital needs for infrastructure replacement and renewal requirements
- Revenue shortfalls due to a population decline, which created significant demands on the utility's Rate Stabilization Fund (created in 1993)

Actions Taken

To address financial pressures, the utility requested a rate increase for the first time in 6 years. PWD needed to raise $19.9 million in revenues in FY 2002, $24.5 million FY 2003, and an incremental $26.2 million in FY 2004. PWD requested rate increases of 6 percent in FY 2002, 6.8 percent in FY 2003, and 6.8 percent in FY 2004.

This request set into motion a review process that was established by regulation in 1992, involving a series of formal and informal public hearings and discovery actions. The review process is presided over by an Independent Hearing Officer and a Public Advocate, both selected by the Mayor, City Council, and City

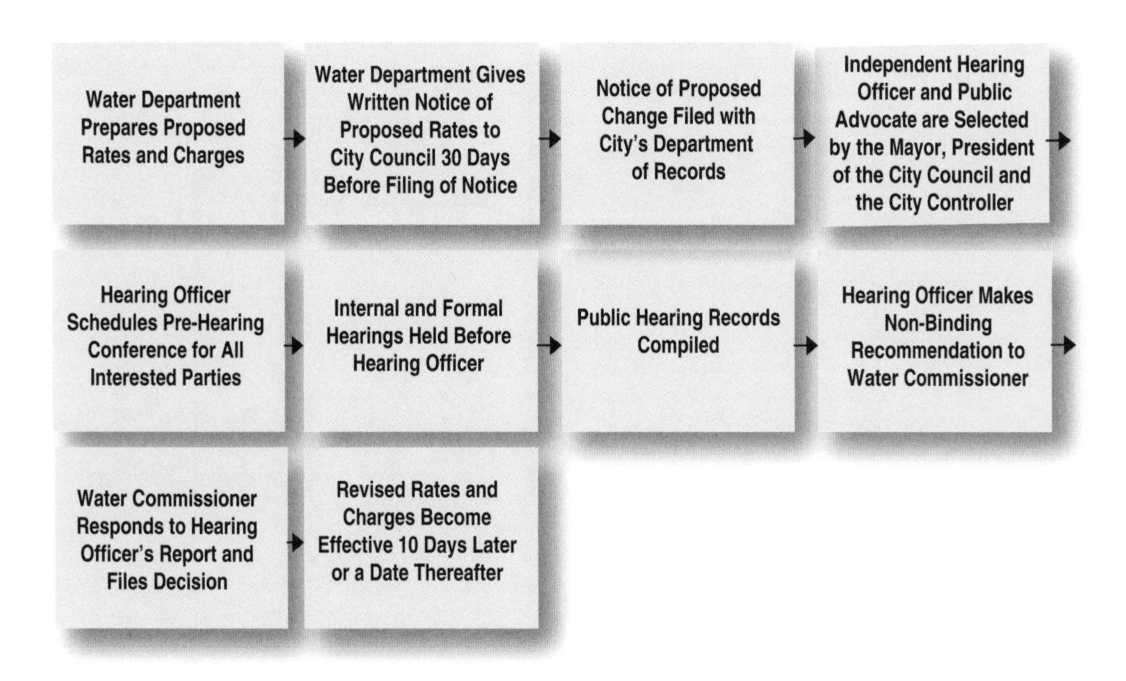

City of Philadelphia Water/Wastewater Rate Setting Process

Controller. The utility funds the services of the Public Advocate (typically a community legal service attorney), at a cost of approximately $200,000 for each review process.

Since this extensive review process was established, PWD has worked with the City and its representatives to shorten its duration to an average of 5 months.

Concurrent with the rate increase, all customers had their serviced charges changed in order to reallocate the costs of stormwater collection and treatment. A Citizens' Advisory Group created in 1995 spent 2 years defining a fair and equitable process for the reallocation of stormwater-related costs.

In addition to the required hearings, PWD communicated to stakeholders through news releases, fact sheets and brochures, website postings, and mailers. The only vocal opposition to the rate increase came from a coalition of condominium owners, who conducted a letter-writing campaign and hired legal representation.

Rate Structure Implemented

PWD's rate increase was approved. The approved rate increase implementation is a multi-year scheme: 5.8 percent in fiscal year 2002, 6.8 percent in fiscal year 2003, and 6.8 percent in fiscal year 2004. Average residential rates rose by just over $3 per month over the 3-year period.

During the review process, PWD was able to explain and justify the rate increase. As a result, "this was the first time in over 30 years that the rates were not taken to court when an increase was requested," said Bernard Brunwasser, Deputy Water Commissioner—Finance and Administration.

Conclusions

"This process confirmed for us the need to make rate increases bite-sized," Brunwasser said. "We had very little comment on rates from most residential customers because the increases were relatively small and came after 6 years of no increases." The utility also used the rate setting process as an opportunity to

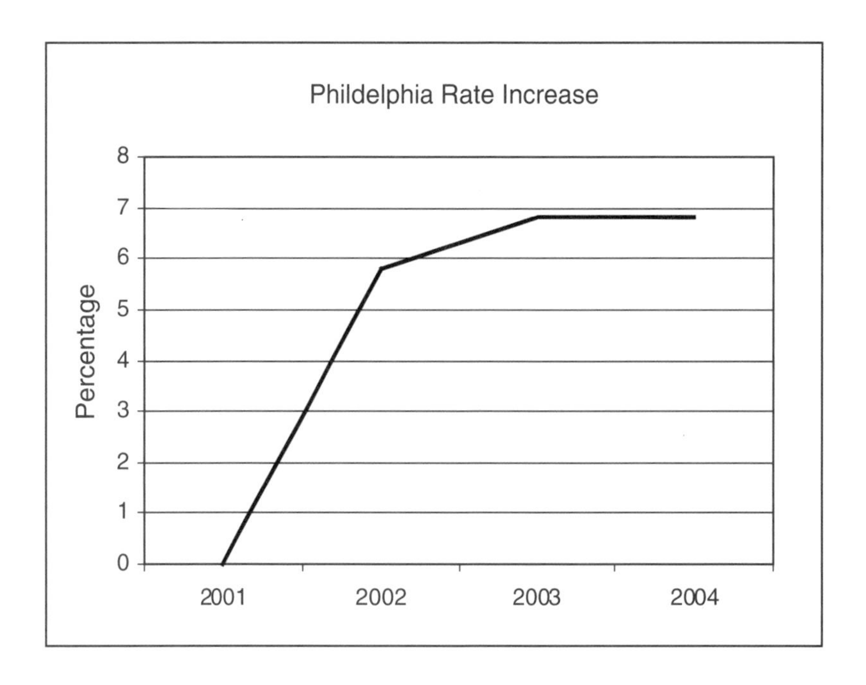

educate elected officials and the community on the need for regular rate increases to avoid rate spikes.

In addition, PWD leveraged the rate increase as an opportunity to communicate positive information. A PWD press release led to a newspaper interview in which the utility communicated its successes, including improvements in river water quality due to source water protection efforts. "We're a forward-looking utility—we try to be ahead of the curve and get the word out so people understand they're getting a good value," said Brunwasser.

PWD is careful to plan conservatively for rate increases. Said Brunwasser, "In our process, data is reviewed with a fine-toothed comb, and you can't go forward with the absolute minimum. You have to be conservative in what you project, and meet your projections. We've done that consistently over the past dozen years or so."

Case Study #9—Portsmouth, VA

Name of Utility	City of Portsmouth Department of Public Utilities, Portsmouth, VA	Median Household Income	$29,400
Population Served	Over 140,000	Number of Connections	31,800
Contact	Jim Spacek—Utility Director, 757.393.8524, spacekj@portsmouthva.gov	Website	www.portsmouth.va.us/publicutil

Lessons Learned

- Demonstrate that the rate increase program will be affordable and will support overall community objectives. This effort can generate approval for increases even in financially strapped communities.
- Publicize successes to enhance utility credibility and build community support for future improvements.

Situation Overview

Founded in 1879, the City of Portsmouth Department of Public Utilities is a self-supporting public service. The Department treats drinking water and manages distribution, wastewater treatment, and sewer lines.

Several influences compelled the utility to pursue a rate increase in the mid-1990s. Utility management was noticing an increase in the costs of emergency response and reactive maintenance, combined with an increase in the number of main breaks due to age. Although the utility had developed infrastructure master plans in 1988 identifying a number of distribution and collection system elements in need of attention, the City Council had not approved funding to address those needs. Management wanted to seize the opportunity to correct a system that was verging on over a 300-year replacement cycle. In addition, regulations mandated upgrades to plant facilities for compliance with Safe Drinking Water Act requirements.

The City of Portsmouth has the fourth lowest household income in Virginia, so raising the capital for the needed and mandated improvements posed a substantial challenge. The City also faced other substantial public investments for schools and revitalization of downtown neighborhoods. Further, the City's tax base is limited, as 40 percent of the property is owned by the federal or Commonwealth governments.

Actions Taken

In 1996, the Department of Public Utilities conducted a strategic financial planning study to fully document its revenue requirements, and to develop rate and financial planning strategies addressing both the system renewal and replacement (R&R) needs and the plant upgrade requirements. The study also included an operations review to identify opportunities for optimizing operation and maintenance. The studies identified an early-implementation strategy, in which R&R efforts would begin in one neighborhood per year, while follow-on valuation and asset management studies were conducted to more fully identify the needs of the system.

In prioritizing neighborhoods for improvement, the Department used GIS data on average system ages by neighborhood (see map below). Operations and maintenance history, such as main breaks and emergency repair needs, also were considered. Finally, the Department considered the ability to coordinate the neighborhood system efforts with overall City Council goals for revitalization.

The rate study report was completed in September of 1997 and recommended a 32 percent increase for water rates and a 79 percent

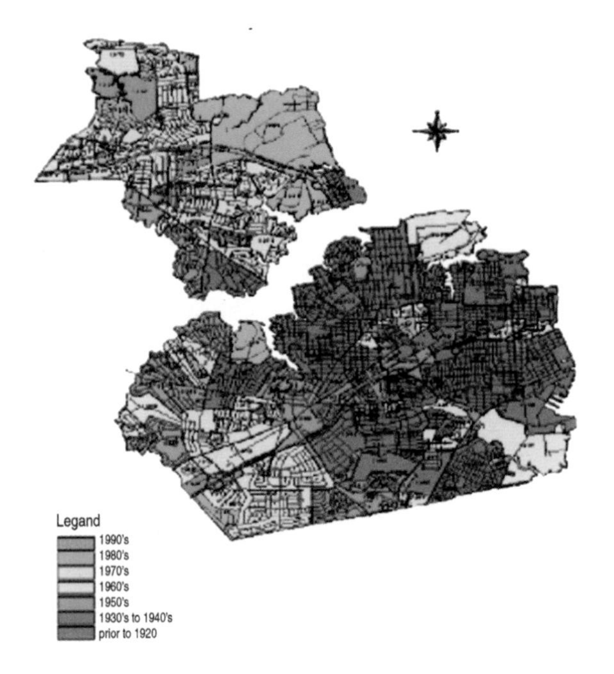

Map showing ages of Portsmouth system assets, by neighborhood

increase for wastewater rates over a 5-year period. The wastewater increase included a reallocation of costs as a result of a separate cost of-service study.

The Department developed a multi-year increase program that would increase funding for R&R in two 5-year phases. In the first phase, the Department needed to increase R&R from less than $500,000 per year to $3 million per year. In the second phase, the Department needed to increase R&R to $9 million per year.

The Department identified several key audiences for the rate increase, including:

- Employees
- Mayor
- City Council
- Utility Board
- Bond rating agencies
- General consumers
- Chamber of Commerce
- Industrial or institutional customers
- Consumer advocacy groups

In addition to the rate study and cost-of-service report mentioned above, the Department developed a number of related technical studies to support the rate increase effort. These included rate impact projections, multi-year financial models, rate comparisons with neighboring communities, asset management and community goals, economic development, and affordability analyses. Given the community's low-income population and the limited amount of taxable property, these studies were instrumental to the rate increase campaign's success.

The Department used a variety of methods for public education/consumer outreach including press briefings, web site postings, mailers, fliers, presentations to City Council and the utility board, public information meetings, public hearings, and brochures. A special briefing of the City Council during its annual retreat provided an opportunity to respond to detailed questions and demonstrate that the multi-year program was affordable.

Rate Structure Implemented

The Department implemented a multi-year rate increase broken into two 5-year phases. The first phase increased water rates by 32 percent and wastewater rates by 76 percent. This increased the typical consumer's monthly bill from $14.50 to $18.50 and wastewater from $6.02 to $10.00.

The Department built its case for the rate increase by structuring its capital program around the City's three stated goals: neighborhood quality, economic development, and fiscal strength.

"While many communities struggle to find the money to replace unseen infrastructure, Portsmouth has shown that this obstacle can be overcome—even in the most difficult circumstances." Virginia Municipal League

Conclusion

The most influential group in approving the rate increase was the Mayor and City Council. Securing their support required strategic planning to fully identify the needs, and demonstrate that the program could be implemented in stages that were affordable and would support overall City goals and objectives. There was no serious organized opposition to the increase.

Election cycles were taken into consideration in passing the initial rate increase program. The Department also considered the timing and amounts of other rate and tax increases in developing a strategy for increasing water rates.

Portsmouth's program has been recognized as a model of responsible leadership in protecting the value of aging utility infrastructure. In September 2002, Portsmouth's asset management and renewal program was awarded the President's Award by the Virginia Municipal League (VML). The President's Award for excellence is the highest award issued by the League. The award recognizes that, despite very difficult economic and financial circumstances, Portsmouth has made substantial investment in renewal and replacement of its aging utilities infrastructure. In granting the award, the VML's judges noted that "One of the most fiscally strapped localities in the State, Portsmouth's government leaders convinced residents to invest in repairs to the collapsing systems beneath the streets, despite a long list of more visible needs across the city... While many communities struggle to find the money to replace unseen infrastructure, Portsmouth has shown that this obstacle can be overcome—even in the most difficult circumstances."

The Department continues to disseminate information on the progress of the improvements and the work for future improvements. These efforts are to help generate support for a continuing efforts. As part of the second 5-year program, the Department is currently in the process of increasing the level of funding for R&R efforts from $3 million per year to $9 million per year, based on the findings of the valuation/asset management studies that were completed in 2001.

Case Study #10—Springfield, MA

Name of Utility	Springfield Water and Sewer Commission, Springfield, MA	Median Household Income (annual)	$30,000
Population Served	■ 252,000 ■ 152,000 Urban (retail) ■ 100,000 Suburban (wholesale)	Number of Connections	43,000 (retail)
Contact	Kathy Pedersen—Director of Public Communications, 413.787.6256 x111, kathy.pedersen@waterandsewer.org	Website	www.waterandsewer.org

Lesson Learned

- Invest in building a reservoir of credibility with the community.
- Tell the truth, answer questions in a straightforward manner, and don't be afraid to say "we have some challenges here, but we're working hard to meet them."
- Develop messages that speak to your stakeholders' concerns—and communicate them consistently and repeatedly.

Situation Overview

Faced with growing infrastructure needs and insufficient funding, the Springfield Water and Sewer Commission (SWSC) crafted a communication strategy to support a 5-year rate increase proposal. A federal consent order, aging infrastructure, security needs, and improved efficiency were initial drivers for the most recent rate increase. The greatest challenge in funding these improvements was that Springfield has a large low-income and fixed-income population.

The most recent significant change to the water rate had been a decrease. The water rate in Springfield had not changed in 10 years when the rate was reduced by $.10. A $.05 increase to the water rate was implemented in 2001. The last sewer rate increase was in 1996 from $1.15 to $1.23 per 100 cubic feet of water used. The community chose to stabilize rates for industrial customers and keep their rates lower than residential rates; however, industrial rates also increased proportionately over the 5-year period.

Actions Taken

The rate increase program was successful in producing revenue to support the capital improvement program, which included repair and replacement of aging infrastructure and actions to meet EPA consent orders for a reduction in discharges from combined sewer overflows. After an internal workshop—which helped focus the SWSC vision, mission, and goals—the Commission conducted interviews with a cross-section of external stakeholders. Business leaders, community activists, wholesale customers and elected officials were interviewed to gauge public perceptions of the Commission, its performance, and receptivity to the proposed rate increase.

After gathering input and data, the Commission crafted messages to clearly communicate the need for the rate increase. "Springfield is not a wealthy community, and any increased burden on its ratepayers had to be carefully justified," said SWSC Director of Public Communications Kathy Pedersen. "But we found out that people thought we were doing a good job, and that they believed their money was going to be well-spent."

A brochure and presentation were prepared to outline the case that:

- infrastructure investments support quality of life;
- management of the system has been improved;
- rates have not kept pace with needs, or even with inflation;

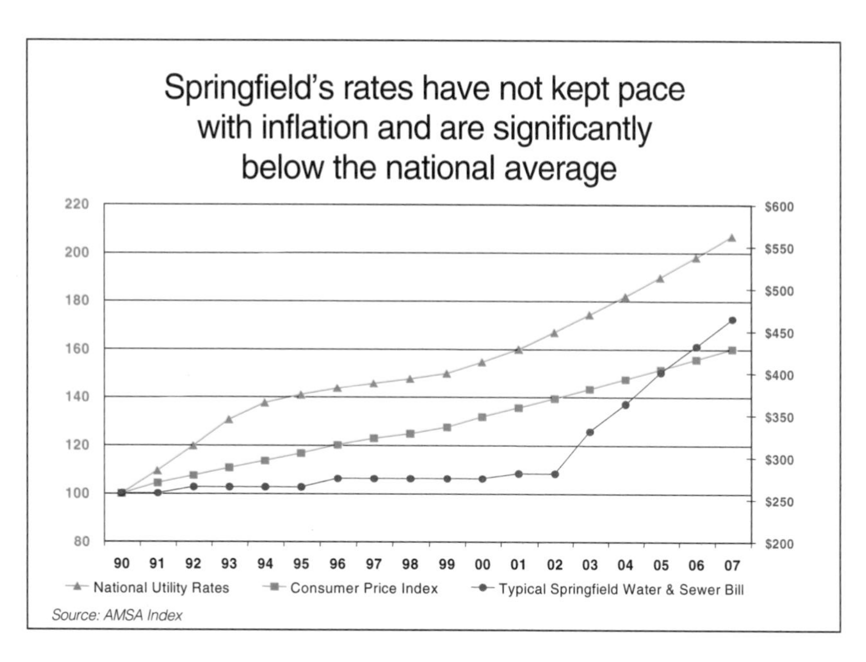

Graphic from a presentation prepared to support SWSC's requested rate increase

- infrastructure investments are needed to improve reliability of the City's underground life support system; and
- specific needs and projects were identified in the 5-year capital plan that would be supported by the rate increase.

The brochure was distributed throughout the community and at Drinking Water Week booths at local malls.

Rate Structure Implemented

After a presentation at a public hearing, the Commissioners voted to increase residential rates from $1.04 per 100 cubic feet to $1.60 by fiscal year 2007, approximately a 54 percent rate increase over 5 years. The increase elevated the typical combined annual water and sewer bill from $275 to $460.

Conclusion

Springfield is in year 2 of the 5-year program and there has been virtually no opposition to the rate increase.

"I believe that the presentation materials and advance preparation of the capital plan helped to educate the public on the need for the improvements and the rate increase to support that work," said SWSC Executive Director Joe Superneau. The money raised will help fund a $70 million capital program of water treatment plant improvements, security upgrades, compliance with federal combined sewer overflow regulations, and rehabilitation/replacement of water and sewer pipes. Without investing in its underground infrastructure, the City anticipated that disruption from almost weekly main breaks would only become worse.

In Springfield's case, a good strategy and clear, effective communication materials were just the tools. The real success lay in the Commission's ability to draw upon a "reservoir" of credibility with the community. They have an experienced, well respected Executive Director and a full-time, high-level staff person in charge of communications. They told the truth, answered questions in a straightforward manner, and were not afraid to say "we have some challenges here, but we're working hard to meet them." That open, honest way of dealing with the public won the day.

Case Study #11—St. Petersburg, FL

Name of Utility	St. Petersburg Water Resources Department, St. Petersburg, FL	Median Household Income (annual)	$37,111 (1999)
Population Served	240,000 residents	Number of Connections	89,000
Contact	Evelyn Rosetti—Program Administration Manager, 727.893.7297	Website	www.stpete.org/pubutil.htm

Lessons Learned

- Present rate increases as part of ongoing programs to improve infrastructure, rather than as unique events to respond to crises.
- Pursue a program of regular annual rate increases. In addition to managing expectations, annual increases can be used to help a utility deal with uncontrollable changes in operational costs and system revenues.

Situation Overview

St. Petersburg, Florida, the largest city in the Pinellas County, provides potable water to its residents and three wholesale water customers. The entire Water Resources Department, through various divisions, provides treatment and distribution of potable water, as well as collection, reclamation, and distribution of treated wastewater for irrigation.

Since the 1930s, the City has relied on groundwater for drinking water supplies. Today, the City purchases water on a wholesale basis from a regional water supplier, which uses a combination of groundwater, surface water, and desalinated water supplies. The regional supplier treats and blends raw water, and the St. Petersburg Water Resources Department continues to operate its own treatment plant to aerate, soften, and filter the incoming water.

Drivers for St. Petersburg's current rate increase program include:

- continuous annual increases in raw water costs;
- declining consumption due to conservation efforts and a wet weather cycle;
- low interest rates; and
- badly needed maintenance for aging infrastructure.

A new rate-setting approach was implemented in 1999, in part to address a strained relationship with some of the utility's wholesale customers. The City has been attempting to negotiate a new wholesale service agreement during the same period as the rate increase program.

Actions Taken

To address the drivers, the Water Resources Department has scheduled a 5-year program of capital improvements to its water and wastewater infrastructure, with costs totaling more than $143 million. As a result, the Department is seeking about $5.3 million per year over the next 5 years in additional funds. This translates into a requested water rate increase of about 15 percent per year for a 5-year period, beginning in 2004.

In the past, approved rate increases typically have been 3 percent to 4 percent lower than requested increases. The reductions in adopted rate increases have been accomplished, in part, by drawing down the utility's unrestricted cash balances. The net effect of these decisions, in combination with the causal factors mentioned earlier, has been that unrestricted cash balances are no longer capable of supporting the revenue shortfall's difference.

In planning for this rate increase program, the Water Resource Department targeted the following groups as critical audiences for public education and outreach:

- Mayor and City Manager
- City Council and Utility Board
- Wholesale customers

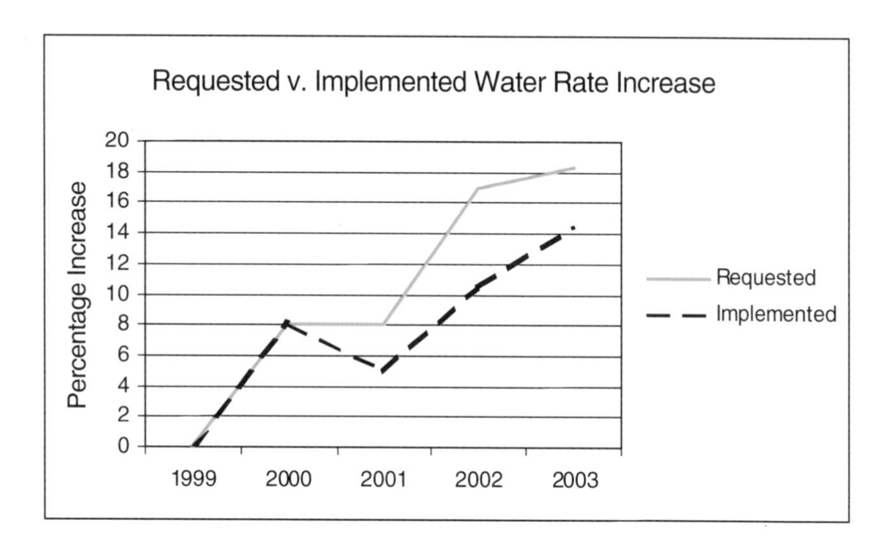

To support its request, the Department conducted numerous studies including rate impact projections, multi-year financial models, rate comparisons with neighboring communities, analysis of historical utility rate increases versus increases in other cost indices, and studies demonstrating the continuing increases in raw water supply costs. The results of these analyses indicated that without significant action, the utility's unrestricted cash reserves would fall below desired levels.

To communicate this information to the public and address stakeholder concerns, the Department has relied on news releases, mailers, fliers, bill stuffers, presentations to the City Council, meetings with wholesale customers, and public hearings. Messages conveyed to the stakeholders included the following:

- The cost of water has increased—and is expected to continue increasing—at a rate of 15 percent or more annually through the near future, per the City's arrangement with the regional water supplier, Tampa Bay Water.
- Billable volumes have declined as a result of conservation and abnormally wet weather conditions; as a result, increasing rate revenue requirements must be spread over fewer units.
- Interest rates at historic lows have significantly reduced interest earnings—a major component of the Department's non-rate revenues.

Rate Structure Implemented

Each year, St. Petersburg updates 5-year projections of its costs and revenues. Rate increases needed to cover the forecast costs are projected for each 5-year period. In general, the utility has sought to equalize the forecast rate increases over each 5-year projection period. However, as the rates are reviewed on an annual basis, only the rates for the subsequent fiscal year are proposed for adoption.

The most recent adopted rate increases raises the typical residential user's annual water bill to $87 more than their 1999 annual water bill.

The proposed rates do not incorporate dedicated fees for infrastructure rehabilitation and repair, but the rate structure does identify the portion of the rates that are intended to recover the utility's costs for raw water supply, over which the utility has no control.

Conclusion

As the Department moves forward with efforts to obtain approval of the needed water rate increases, many variables still are under consideration. These variables include the potential for accompanying increases to sewer rates, potential changes to a combination of the causal factors that are driving the current rate increase program, and consideration of other utility rate increases proposed during the same time period.

At the time this case study was prepared, there was no organized opposition to the rate increase. The rate increase program was not presented as a unique event, but rather as part of an ongoing improvement program to infrastructure, and response to other uncontrollable changes in operational costs, water usage, and other revenues.

Case Study #12—Tucson, AZ

Name of Utility	Tucson Water, Tucson, AZ	Median Household Income (annual)	City of Tucson: $30,981 Pima County: $36,758
Population Served	680,000; serves 30% outside Tucson proper	Number of Connections	211,000
Contact	Barbara Buus—Manager of Rates and Revenues, 520.791.2666, bbuus1@ci.tucson.az.us	Website	www.ci.tucson.az.us/water/

Lessons Learned

- Be thorough in identifying stakeholder groups and developing an understanding of their concerns. Leaving critical stakeholder groups out of the outreach process can delay or derail a rate increase.
- Keep water rate increases as low as possible—but keep them coming.
- Consider the feasibility of implementing capital recovery fees, such as system equity fees, to help reduce the level of required rate increases—but make sure you consider and address the effects of these fees on all stakeholder groups.

Situation Overview

Tucson Water is owned by the City of Tucson. The governing body of the City consists of a Mayor and six Council members; the governing body is elected City-wide. All major financial activities of Tucson Water, including changing water rates and fees, require the approval of the Mayor and Council.

Tucson Water is the major water provider in Pima County, Arizona, serving both inside and outside the corporate limits of the City of Tucson. The utility operates a potable water system and a reclaimed water system, the latter for turf irrigation. Water rates and fees are the same inside and outside the City limits. Median household income (1999 from 2000 Census data) for the area served varies greatly between the City and the County: City of Tucson, $30,981; Pima County, $36,758.

Tucson Water is self-supporting from its rates and fees. Historically, the utility's major funding sources have been water sales revenues, miscellaneous service fees (water service turn-on charges, bill delinquency fees, water service installation charges, fees for providing billing services to other entities), and interest earnings on water-generated revenues.

Tucson Water provides an interesting historical perspective on water rate increases. From 1976 through 1992, the governing body approved annual rate increases for the utility; the average annual increase approved was 7.2 percent. A major factor contributing to this solid increase history was a crisis. In the early 1970s, the utility had not raised rates for several years, resulting in the need for a 30 percent rate increase in 1974. That increase was approved by the governing body—but many constituent complaints resulted, related both to the level of the increase and to certain rate elements adopted with the increase. Ultimately, several members of the governing body were recalled. City and utility management, as well as the new governing body, learned from the crisis: keep water rate increases as low as possible but keep them coming.

Another crisis, however, was a major contributor to breaking the 17-year history of annual water rate increases. In the early 1990s, Tucson Water shifted from groundwater to river water (Central Arizona Project) as its potable water supply. Although the shift had been planned for over 20 years and a massive public outreach effort related to the change had been undertaken, difficulties with "red water" occurred during implementation. The Utility lost credibility; water rate increases were impossible to request during this period and would have been impossible to get. Tucson Water emerged from this crisis slowly, a critical factor in the emergence being returning directly to customers to find out what quality of water they expected "at the tap" and what options were acceptable to customers for achieving that quality. The ability to make a case for water rate increases and receive such increases returned as credibility was restored so that by the late 1990s the utility was receiving rate increases of around 3 percent every 2 years.

Actions Taken

Financial Planning: Water Rates

Tucson Water conducts an annual financial planning process whose first major product is a six-year Financial Plan. (The current budget year is year 1; year 2 is the "test year" for a water rate change; years 3 through 6 represent projection of requirements and financing of those requirements.)

In recent years, based on the Financial Plan, annual water rate (water sales revenue) increases of around 4 percent have been sought and adopted by the governing body. The process to achieve increases at this level has involved at least two key undertakings:

1. "early warning" to the governing body of the level of revenue increase needed; and
2. a two-pronged approach to customer/constituent input.

Early warning to the governing body has been achieved by:

- one-on-one meetings with each member of the governing body to discuss the Financial Plan and its conclusions; and
- presentation of the Financial Plan to the entire governing body prior to the governing body's review of proposed budgets.

The latter allows the governing body to reduce the proposed budgets, thereby reducing revenue requirements and the level of revenue increase in the Financial Plan, if the level of revenue increase is deemed unacceptable. Alternatively, the governing body adopts the Financial Plan, thereby agreeing to the level of revenue increase in the Plan and paving the way to designing new water rates to generate the revenue targeted in the Plan.

The two-pronged approach to customer/constituent input involves both the Financial Plan and rate design. The **Citizens' Water Advisory Committee,** in existence since 1977, reviews Tucson Water's revenue requirements and resulting Financial Plan and then recommends (or not) approval of the Plan to the governing body. This Committee consists of 15 unpaid members, seven appointed by members of the governing body and eight appointed by Tucson's City Manager. About half the members of the Committee reside outside the City limits.

Following the adoption of the Financial Plan, which establishes the amount of revenue required from water sales, a **Customer Rate Design Group** convenes. This Group considers how the revenue required should be allocated to Tucson Water's different customer classes and what rate structure/format (for example, inclining block) seems most appropriate for each customer class. The Group then presents its recommendations to the governing body. Typically, the Group consists of 10 members, each customer class being represented by at least one member. Neighborhood associations, both inside and outside the City limits, and professional organizations with an interest in water matters, such as the Chamber of Commerce, recommend members for the Group.

Financial Plan: New Capital Recovery Fee

In May 2002, the governing body adopted a Financial Plan calling for a 4.3 percent increase in water sales revenue to fund the requirements of Fiscal Year 2003, which began July 1, 2002. (Water rate changes to generate this increase became effective in October 2002.) In his presentation of the Plan to the governing body, the Water Director indicated that staff would be reviewing the Utility's costs currently funded by water sales revenues to determine if another cost-recovery source would be more appropriate.

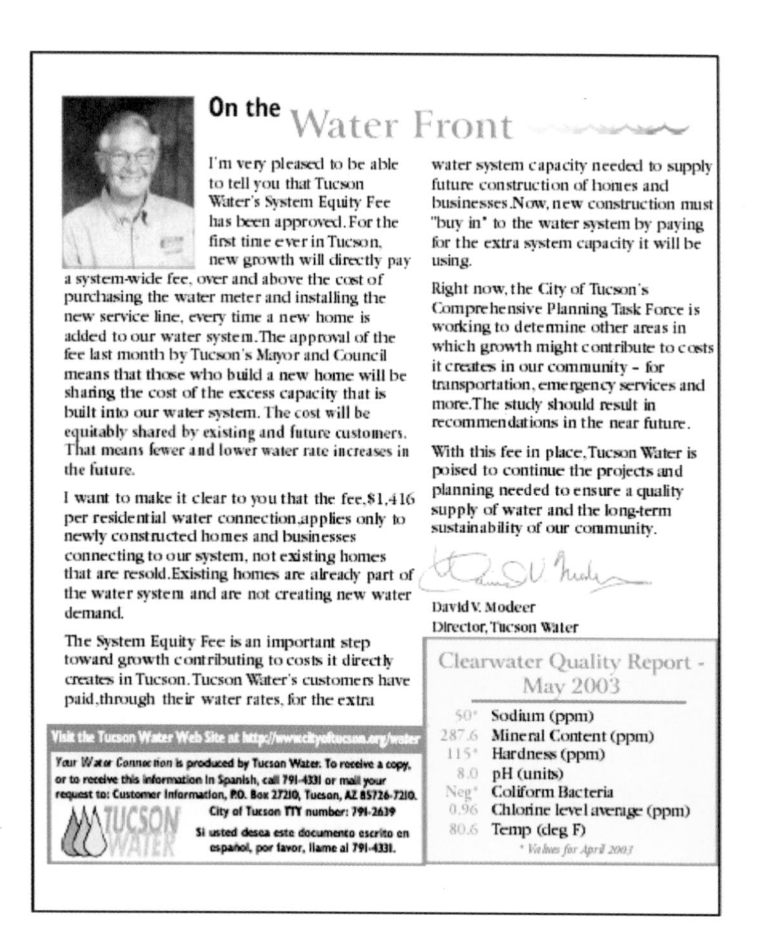

On the Water Front

I'm very pleased to be able to tell you that Tucson Water's System Equity Fee has been approved. For the first time ever in Tucson, new growth will directly pay a system-wide fee, over and above the cost of purchasing the water meter and installing the new service line, every time a new home is added to our water system. The approval of the fee last month by Tucson's Mayor and Council means that those who build a new home will be sharing the cost of the excess capacity that is built into our water system. The cost will be equitably shared by existing and future customers. That means fewer and lower water rate increases in the future.

I want to make it clear to you that the fee, $1,416 per residential water connection, applies only to newly constructed homes and businesses connecting to our system, not existing homes that are resold. Existing homes are already part of the water system and are not creating new water demand.

The System Equity Fee is an important step toward growth contributing to costs it directly creates in Tucson. Tucson Water's customers have paid, through their water rates, for the extra water system capacity needed to supply future construction of homes and businesses. Now, new construction must "buy in" to the water system by paying for the extra system capacity it will be using.

Right now, the City of Tucson's Comprehensive Planning Task Force is working to determine other areas in which growth might contribute to costs it creates in our community - for transportation, emergency services and more. The study should result in recommendations in the near future.

With this fee in place, Tucson Water is poised to continue the projects and planning needed to ensure a quality supply of water and the long-term sustainability of our community.

David V. Modeer
Director, Tucson Water

Visit the Tucson Water Web Site at http://www.cityoftucson.org/water

Your Water Connection is produced by Tucson Water. To receive a copy, or to receive this information in Spanish, call 791-4331 or mail your request to: Customer Information, P.O. Box 27210, Tucson, AZ 85726-7210.

TUCSON WATER

City of Tucson TTY number: 791-2639

Si usted desea este documento escrito en español, por favor, llame al 791-4331.

Clearwater Quality Report - May 2003

50*	Sodium (ppm)
287.6	Mineral Content (ppm)
115*	Hardness (ppm)
8.0	pH (units)
Neg*	Coliform Bacteria
0.96	Chlorine level average (ppm)
80.6	Temp (deg F)

* Values for April 2003

Tucson Water newsletter announcing approval of the system equity fee

This review resulted in a new fee called the **'system equity' fee**. The rationale behind the fee follows: Capital improvements to provide additional water system capacity have been constructed in advance of when the additional capacity will be fully utilized; current system users have funded these improvements via water rates and have, therefore, provided system capacity to serve future users. By paying the one-time system equity fee when connecting to the water system, a new user 'buys in,' funding the new user's proportionate share of system capacity.

The revenues generated by the fee, used to pay debt service, reduce the level of increase in water sales revenue which would otherwise be required. The next annual update of the Financial Plan in 2003, which included revenues from the system equity fee beginning in FY 2004, revealed that the annual water sales revenue increase required was less than 1 percent and the biennial increase about 2 percent. In June 2003, the governing body adopted a Financial Plan which included the system equity fee beginning in FY 2004 and biennial increases in water sales revenue, the first increase scheduled for FY 2005 (October 2004).

The system equity fee had to be developed and marketed within a very compressed time frame since the goal was to have the fee replace, in effect, the water rate increase needed for FY 2004; in addition, approximately 6 months of the time available had to be allocated for the legal process to adopt a fee of this type mandated by Arizona State law. Thus, there was no time for public input into the basis, or calculation, of the fee; rather, outreach efforts were directed to explaining the rationale behind the fee and how the calculation had been made. Presentations were made to the local homebuilders' association and related groups, a workshop for developers was conducted, and a public meeting open to all concerned citizens was held. These efforts were in addition to the public education and outreach efforts represented by meetings with the Citizens' Water Advisory Committee, the Customer Rate Design Group, and members of the governing body.

Rate Structure Implemented

Outreach efforts initially appeared successful. In January 2003 the legal process to implement the fee was initiated; the governing body unanimously adopted the notice of intention to adopt the fee and scheduled a public hearing for April. The public hearing revealed that one major interest group had been overlooked in the outreach efforts: non-profit developers. As providers of affordable housing—affordable housing already known to be a significant concern in the Tucson area—these developers made a strong case that the system equity fee would seriously affect the amount of such housing they would be able to construct. Tucson Water began to work on a solution with these developers and the City's Community Services Department, which oversees funding for affordable housing. The governing body rejected the fee in its first vote following the public hearing but later voted to adopt the fee when the non-profit developers indicated their issue had been resolved.

Conclusion

The utility's initial failure to include the non-profit developers in its outreach efforts helped lead to a one-month deferral in the scheduled implementation date of the fee and concomitant loss of revenue. The better news is that it is a one-time revenue loss, and the fee is expected to fulfill its intended role in future years.

APPENDIX B—"AVOIDING RATE SHOCK" TOOLKIT

Contents

This appendix includes the following sections:

- Rate structure options and technical studies
- Example materials:
  - Brochures
  - Fact sheets and Newsletters
  - Presentations

Rate structure options and technical studies

Considerable attention has been devoted to identifying and evaluating rate structure options during the past several years. For example, in 2000, AWWA released an updated edition of its "M1" rate manual[1], which devoted more than twenty chapters to descriptions and discussions of various rate structure and fee options. These options range from traditional rate structures, such as uniform rates and declining/inclining block rates to emerging rate options, such as negotiated rates and miscellaneous and special charges.

The current study is not intended to supersede the body of rate study work represented by the update to the "M1" manual and related industry documents. Rather, it is one objective of this report to highlight some of the rate increase options and related technical studies from the options identified in the rate literature. These options often merit consideration when a utility is faced with implementing a considerable rate increase. The reader is advised to consult additional works, such as the 2000 edition of the "M1" manual, for more details related to these rate structure and study options.

Rate structure options and policy approaches that are often found to be worthy of consideration for utilities faced with substantial rate increases include:

- **Special fees and charges (service connection fees, impact fees, other ways to ensure user of those services pays)**—For example, utility systems that don't currently charge an impact/system development charge might consider documenting appropriate amounts that could be charged and imposing such fees, which would help reduce the pressure for general rate increases. Or, if an existing impact fee/system development charge exists but does not fully recover identified related costs, these fees might be increased to more fully recover identified allowable charges.
- **Special dedicated fee or rate component for renewal and replacement**—There may be value in identifying the rate component that is related to addressing renewal and replacement (R&R) needs that are documented through asset management studies or other means. This may be useful in making the case to governing boards and stakeholders that the additional fees represent value added—providing a higher level of service than had been provided earlier. In some cases, this might take the form of an identified portion of a volumetric rate (e.g., $0.50/1,000 gallons of the $3.00/1,000 gallon rate is related to addressing $5 million in renewal and replacement needs that have been identified through detailed asset inventory and condition assessment efforts). In other cases, this might take the form of a separate charge added onto the bill to address R&R needs.
- **Increase in base charge**—Many utilities do not fully recover even the costs of billing and customer services through their monthly or quarterly base charges. When faced with major increases in revenue requirements, it is often an appropriate juncture to realign these charges to fully cover customer and billing functions, and possibly to cover some portions of other fixed costs of the system.
- **Enhanced/revised customer classification**—Substantial increases in revenue requirements

1. *Principles of Water Rates, Fees, and Charges*: AWWA Manual M1—Fifth Edition, American Water Works Association, 2000.

may serve as a trigger to revise a utility system's customer classification system. For example, it may become warranted to separate out certain industries whose water demands (either in time or quality) require special treatment into a new rate class or category, as the dollars involved become sufficient to justify the added administrative expense of maintaining additional required records. Such efforts may be important both in helping to raise required revenues and in documenting to stakeholders that customers are paying in proportion to their demands on the system.

- **"Lifeline" rates or discounts to protect elderly or other special groups**—When faced with a major increase in revenue requirements, it is often useful to consider implementing or revisiting programs that protect special constituencies within the utility's customer base (e.g., elderly customers, low income customers). Overall support for the required rate increases may hinge on being able to respond to questions from governing board members or other stakeholders that the needs of these special groups have been appropriately addressed. If no programs for special groups exist, it may be an appropriate juncture to consider development of such a program. If an existing program exists, it may be important to revisit the program, to ensure that the special groups are treated appropriately in light of the increased revenue/rate requirements.
- **Multi-year rate program**—Some utilities that have traditionally adopted a budget and rate for only a single year have found it useful to project and adopt multi-year rate increases when faced with major changes in revenue requirements. For example, the Alexandria Sanitation Authority, one of the case study utilities for this study, adopted a five-year rate increase when faced with doubling of rates to meet regulatory requirements. Adopting a multi-year program provides customers an element of predictability. It also provides a basis for getting decision-maker and stakeholder buy-in to the long-term impact of required rate adjustments.
- **"Automatic" rate increases**—Another variant on the planned multi-year rate increase is an automatic rate increase, in which a utility adopts a 'standing' rate increase that remains in effect unless removed. One example of this is the case of Orange County (Florida) Utilities (OCU). In the late 1980s, OCU adopted an automatic annual increase for water and wastewater rates. This came after much difficulty was encountered in implementing substantial required rate increases in the mid-1980s. OCU's Board adopted a resolution stating that on October 1st of each year (start of fiscal year), water and wastewater rates would increase by 3% as long as no other action was taken. So each October bill had the new rate with a statement on the bill reminding the customer of the "automatic" increase. After a few years, the utility's revenues were sufficient, and the Director was able to go to the Board some years (usually politically timed) with a request to formally waive the automatic increase for the coming year. Of course, this was greeted with great enthusiasm and gratitude for the Utility, and provided much goodwill from the customers in the years in which the increase could be waived. By creating the expectation that normal, inflationary increases should be expected, the utility was able to keep up with rising costs and program needs.

Technical studies and communication support efforts that are often found to be worthy of

> *"Whereas the use of capital and other funds for avoiding needed rate increases detracts from the utility's long-term stability, the use of specifically designated rate stabilization funds (generally resulting from surplus moneys) to reduce (but not eliminate) an atypically large rate increase, may benefit a utility...When future financial needs are known, multi-year rate approvals are another tool to protect rate-setting from outside influence. If rates are regularly adjusted, then incremental increases should be smaller and thus easier to approve—even as a whole."*
>
> —James Wiemken, Director, Standard & Poors Credit Market Services

consideration for utilities faced with substantial rate increases include:

- Cost-of-service studies to document revenue requirements and where they should be allocated within the utility's customer base
- Asset management studies that document revenues required to maintain the system's assets at a target condition or performance level, and/or cost tradeoffs and efficiencies that result from increased R&R expenditures
- Rate impact projection
- Multi-year financial models and scenario analyses
- Comparisons with neighboring communities
- Demonstration of connection between sound utility assets and accomplishing other community goals, such as economic development
- Affordability analyses that compare projected charges with household income and other affordability metrics
- Analysis of historical utility rate increases vs. increase in CPI, CCI, and other cost indexes
- Pictures or sample sections of tuberculated pipes or other examples of deteriorating system elements
- Maps showing the age of pipes in various districts and neighborhoods

Comparing Utility Rates to Other Expenses

In comparing utility rates to other expenses, it is recommended that utilities avoid examples involving discretionary costs (e.g., cable television, movie rental). Here is an example of what *not* to do in comparing utility costs to other expenses:

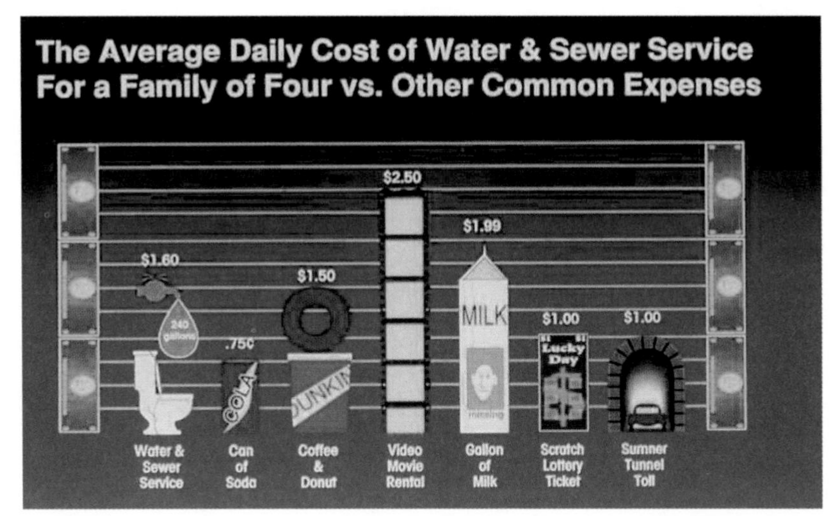

A more effective approach is a simple comparison of utility rates to other utility rates from neighboring communities, as in this example:

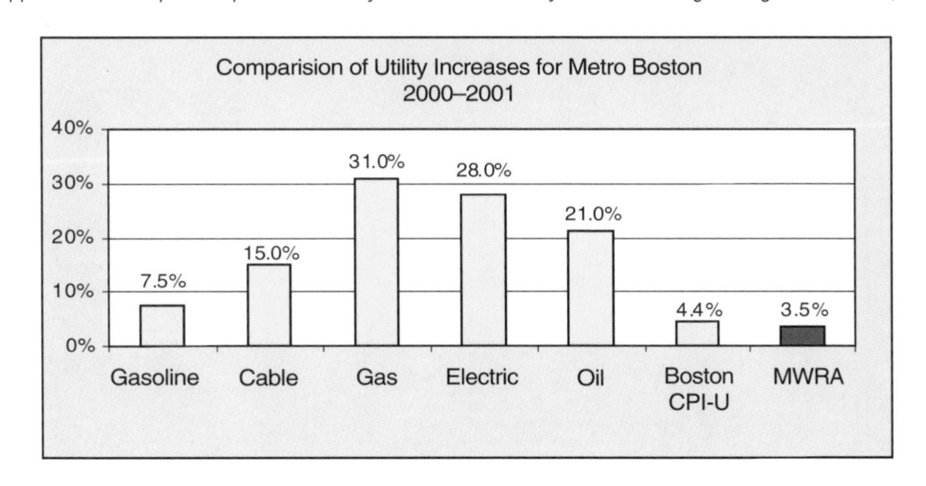

BROCHURES

EVERY day, 24/7, 365 days a year, Las Virgenes Municipal Water District is treating wastewater from more than 85,000 local residents. Our service is transparent, safe, environmentally friendly and energy efficient — and that's the way it should be.

Costs to provide service, however, have nearly doubled in the last 10 years—with skyrocketing energy, environmental, security and insurance prices. Cost containment efforts are no longer enough, making it necessary to raise customer charges.

FAST FACTS ABOUT SEWER SERVICE

- 15,958 sewer connections in the LVMWD service area
- 56 miles of trunk sewer lines operating 24/7
- 9 million gallons of wastewater safely treated each day

SEWER SERVICE CHARGES

WHAT ARE

Other Resources

For information about your water and sewer bill and payment assistance programs, call Monday through Friday, 7 a.m. to 7 p.m. 215-686-6880

To order a copy of "Know Your Rights as a Residential Water and Sewer Customer" 215-686-6880

All residential properties must have an automatic meter reading device. If an automatic meter reading device has not been installed in your property, please call. 215-685-6300

For water and sewer emergencies call 24 hours a day 215-685-6300

Payments and Customer Service Locations

DEPARTMENT OF ENVIRONMENTAL PROTECTION
MONTGOMERY COUNTY, MARYLAND

The Montgomery County Water Quality Protection Charge
Frequently Asked Questions

Rain and stormwater runoff is an issue few people spend much time worrying about unless they have forgotten an umbrella during a downpour or come home to a flooded basement. However, taking appropriate steps to control stormwater runoff is becoming an extremely important issue for Montgomery County.

Impervious surfaces such as roofs, driveways, parking lots, and streets prevent precipitation from entering the ground and the groundwater where it completes the hydrologic cycle. Instead stormwater is collected and either sent to a stormwater facility or discharged directly to the streams without control. Older, urbanized areas of the county without stormwater controls bear witness to the devastation visited upon nearby stream valleys, which were blasted by incredible volumes of water, sediment, and pollution, changing from gurgling, bucolic streams to 50 foot wide lifeless channels with toppled trees, exposed sewer lines, and deeply cut and eroded banks.

What is the Water Quality Protection Charge (WQPC)?

Who will pay?

What are associated nonresidential properties?

How has the charge been determined?

Securing the Future
Investing in Our Infrastructure

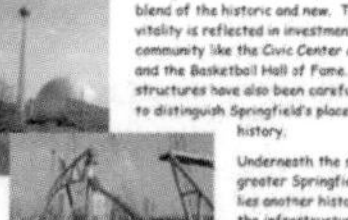

"The Springfield Water and Sewer Commission improves the quality of life for our community through public health protection, environmental stewardship and support of sustainable economic development."
SWSC Vision Statement

How do infrastructure investments support our quality of life?

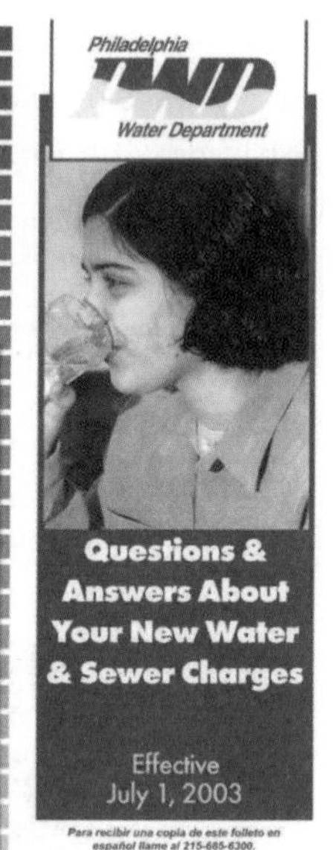

Questions & Answers About Your New Water & Sewer Charges

Effective July 1, 2003

Securing the Future
Investing in Our Infrastructure

Portsmouth is a city that celebrates its history. Evidence of Portsmouth's role in America's past is found everywhere—in its carefully restored architecture, the seaport, and the cobblestone streets. But there's more to Portsmouth's history than meets the eye. Construction of a huge underground infrastructure network for water and wastewater service began in the late 1800s. Today, 550 miles of water distribution piping still provide reliable, safe drinking water from the City's water treatment system, and another 440 miles of sewer conveyance piping transports wastewater to the Hampton Roads Sanitation District for treatment. But the system's reliability for future operation is in question.

The Portsmouth DPU is one of the state's largest public utilities, with approximately 550 miles of water pipe, 31,000 meters, and 210,000 billings producing over $20 million in revenue per year.

Had City residents not renovated the Commodore Theatre and the Victorian-era homes dotting our neighborhoods, these structures would be in a state of evident distress—and that is the status of our underground infrastructure. Because of a historic low rate of spending on infrastructure replacement, we now face a substantial financial challenge. An extensive investment in our infrastructure will be required to preserve the quality of life that has been established aboveground. In the City today, nearly 300 sewer collapses wait to be addressed, and our water pipelines frequently experience breaks. Habitually responding to a growing incidence of emergencies is a costly—and risky—way to manage our infrastructure.

While many visible historic structures have been restored, our underground systems have continued to age. Two thirds of Portsmouth's water pipelines are now clogged with rust and debris.

Portsmouth is not alone in this crisis. A recent report by the American Water Works Association (AWWA) on the nation's drinking water infrastructure found that spending on pipe replacement must **triple over the next 30 years**—representing an additional $250 billion—to maintain safe, reliable drinking water infrastructure. And putting off the investment only increases the eventual cost.

> *"We have used all the funds allowed for maintenance, but each year greater deterioration is occurring..."*
>
> ***X.D. Murden, Portsmouth Superintendent of Water Annual Report—1967–68***

How do infrastructure investments support the City's mission?

The City Council has articulated Portsmouth's mission for the future as a three-part objective: economic development, fiscal strength, and neighborhood quality. Replacing and repairing our failing utility infrastructure directly supports each of these elements:

- it facilitates **economic development** by providing services for redevelopment, business expansion, and new growth;
- it promotes **fiscal strength** through more efficient system operation and a controlled investment program that minimize burdens on ratepayers; and
- it protects **neighborhood quality** by minimizing flooding and street cave-ins and continuing the City's history of safe, reliable service.

Realizing the City's three-part mission requires investing in a sound infrastructure system.

What have we done to begin addressing our infrastructure needs?

In 1996, the City began a strategic financial planning process that was aimed at addressing the existing backlog of utility renewal and replacement needs:

- In December 1997, the City Council approved a referendum to fund $25 million worth of improvements identified in the strategic financial plan, including $13.4 million for neighborhood projects.
- In May 1998, 77% of Portsmouth's voters approved this important start to the renewal program.
- In 1999, work began in the first of the neighborhoods (Park View).
- An asset valuation study recommended in the 1996 strategic plan has just been completed. It provides more accurate information on the value of Portsmouth's utility system and its true renewal and replacement needs.

News Coverage Has Helped Raise Awareness of Infrastructure Needs

MARTIN SMITH-RODDEN/VA PILOT

Albert J. Williams Jr., a senior engineer with the City, holds a 100 year old pipe representative of many older parts of the City.

A June 21, 1999, article in the Virginian-Pilot *publicized the start of the rehabilitation program in Park View, the first Portsmouth neighborbhood to be addressed. The photo and caption shown at the left were part of the article.*

Through the Neighborhood Water & Sewer Replacement Program, the City began to renew water and sewer lines on a large scale. Part of the voter-approved 1998 Utilities Bond Referendum, this program was designed to systematically replace or rehabilitate water and sewer lines throughout the city, one neighborhood at a time. Much of what is being replaced is 2", 6" and 8" pipe, ranging in age from 45 to 100 years old. In its place, the City is installing pipes that are corrosion-resistant and have extended life expectancies. Where possible, a folded lining system is being installed in existing wastewater piping systems to reduce replacement costs and avoid disruptive open-cut construction.

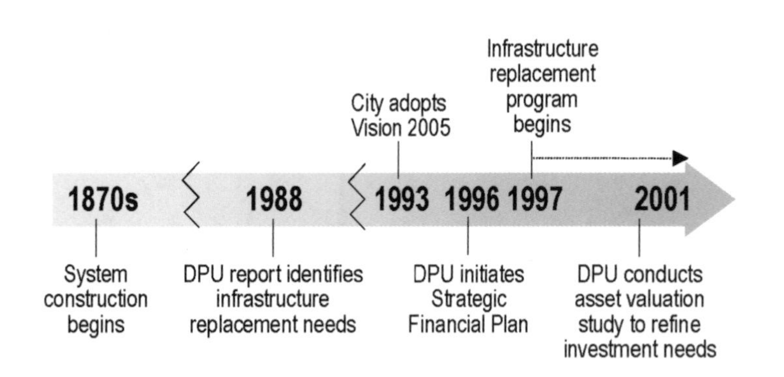

Infrastructure replacement began in earnest in 1997, with a program to address one neighborhood per year. More recent study shows that a more aggressive renewal and replacement program is needed to maintain reliable services, based on industry standards for cost-efficient use of utility system assets.

A folded lining system for sewers and a regular program of line flushing for water are two economical approaches the DPU has taken to extend pipeline life.

What is the investment required to secure our future?

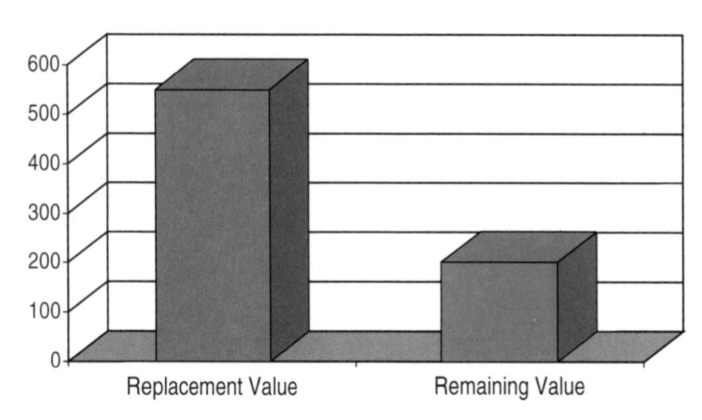

As shown in the chart above, Portsmouth's utility system has a replacement value of about $550 million. However, almost two thirds of the system's value is currently depreciated, using typical useful-life factors.

- Portsmouth's utility system has a replacement value of about $550 million. Based on industry standards, Portsmouth should be spending about $9 million per year to renew and replace the aging portions of the ***entire*** utility system.
- About $6 million of the $9 million per year total is needed to repair and replace aging water and sewer lines in the City's *neighborhoods*, many of which are more than 100 years old.
- Because Portsmouth, like most cities in the United States, has historically not provided adequate funding for renewal and replacement of utility assets, additional funding will be needed to address a backlog of repair and replacement needs.
- Proactive renewal and replacement saves money in the long run. Targeted renewal is much less costly than emergency replacement when facilities fail.

What are the next steps?

Securing the Portsmouth utility system's future requires support from voters, customers, City staff leadership, and the City's elected officials. Critical action items to be undertaken in the next year include:

- Expand stakeholder education/outreach programs to increase awareness of the needs of the utility system and secure support for needed funds.
- Approve the funding for capital renewal of neighborhood water and sewer lines, building to a level of $6.4 million per year as proposed in the FY 2003-2008 Capital Improvement Program.
- Approve renewal funding in the FY 2003 operating and CIP budgets for other components of the DPU system, such as treatment facilities, based on results of the valuation study.
- Determine the appropriate mix of rate revenues, transfers, and other funding sources as part of the DPU's FY 2003 budget cycle.
- Move forward to conduct a condition assessment study to identify and prioritize deficiencies that result from historical funding. Adjust the financial plan, as appropriate, to address these needs.
- Authorize a bond issue for the next leg of the DPU's capital program, including renewal and replacement needs.

Annual Expenditures for Renewal of Neighborhood Facilities

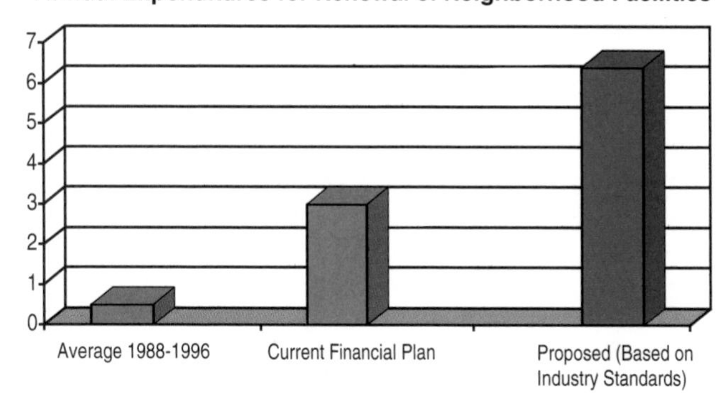

Based on utility industry standards for annual capital expenditures on system renewal and replacement, the City of Portsmouth should be spending approximately $6.4 million per year on renewal and replacement of its ***neighborhood*** *water and sewer lines—more than twice the amount included in the current budget and financial plan. The neighborhood program represents 2/3 of the $9 million in annual renewal needs of the entire system.*

> *"Pipes are expensive, but invisible. Pipes are hearty, but ultimately mortal. Pipe replacement needs are a demographic echo. Increased expenditures are needed to climb the ramp and avoid a gap... The bills are now coming due, and they loom large."*
>
> "Dawn of the Replacement Era: Reinvesting in Drinking Water Infrastructure," American Water Works Association, 2001

What has been the cost thus far?

The figures at right show the increases in water and sewer bills for a typical residential customer during the past five years. These typical charges are compared with the rates projected in the financial plan endorsed by the City Council in 1997. The guiding principles that were used in establishing rate increases included:

- Have gradual rate increases to cushion the impact.
- Fund some smaller capital projects with cash instead of more costly debt. This helps maintain a strong bond rating and favorable interest rates when bonds are issued.
- Use some transfers from the City's General Fund to cushion the need for rate increases.

What are future costs expected to be?

Additional rate increases will be needed to support the renewal program that we now know is required. As shown in the bottom figure, even with projected rate increases, our utility bills will stay well within the affordability guidelines established by the U.S. Environmental Protection Agency. To make sure the program is affordable for utility customers during the next five years, the City will continue to follow the guiding principles identified above. In addition, the City will:

- Aggressively pursue outside funding to reduce rate impacts on Portsmouth's residents and businesses. For example, in 2001, the City secured a $2.6 million grant from the Economic Development Administration to help relocate the Lake Kilby Electrical Building, which was flooded by Hurricane Floyd.
- Refinance debt when lower interest rates create opportunities to save money.
- Monitor utility rates in neighboring jurisdictions to ensure that Portsmouth's rates remain competitive.

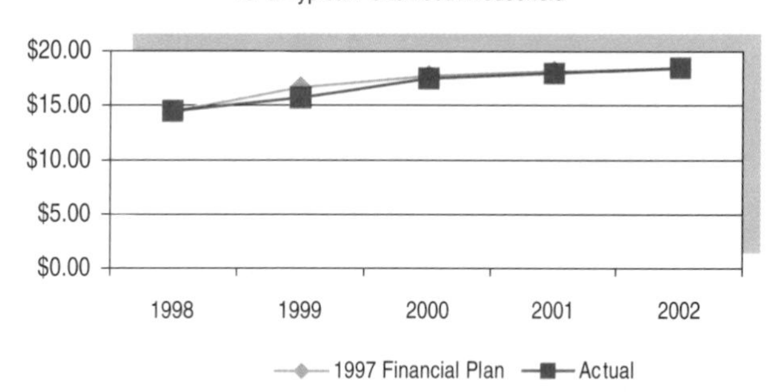

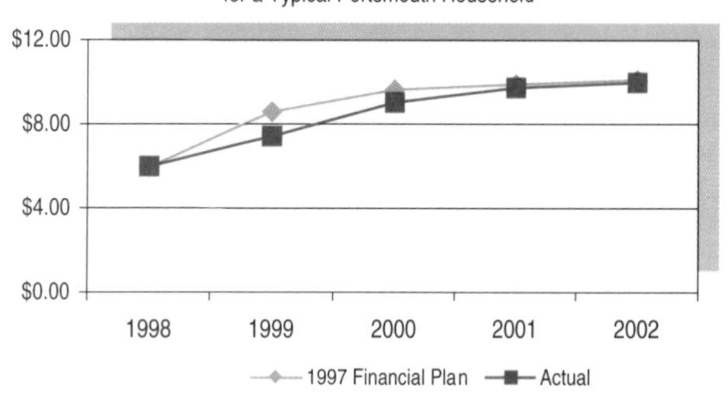

The rates and charges adopted during the past five years have been used to fund the ongoing operation of Portsmouth's utility systems, plus the renewal and replacement of aging water and sewer lines in one average-sized neighborhood per year.

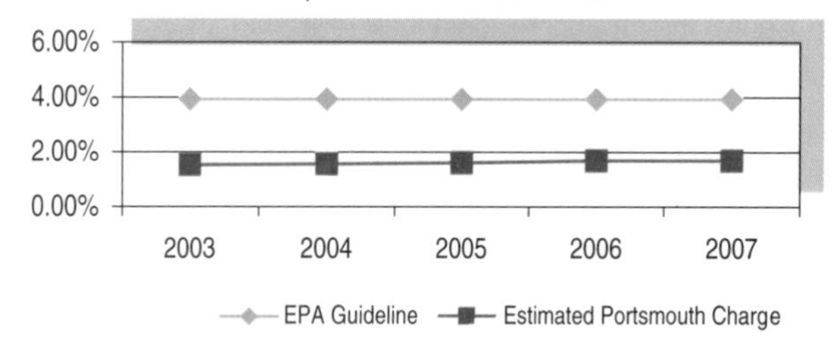

With the rate increases needed to support the renewal program over the next five years, our water and sewerage charges will comprise about 2% of the income for a typical Portsmouth household. This is about half of the 4% affordability guideline established by EPA.

DEPARTMENT OF ENVIRONMENTAL PROTECTION
MONTGOMERY COUNTY, MARYLAND

The Montgomery County Water Quality Protection Charge

Frequently Asked Questions

Rain and stormwater runoff is an issue few people spend much time worrying about unless they have forgotten an umbrella during a downpour or come home to a flooded basement. However, taking appropriate steps to control stormwater runoff is becoming an extremely important issue for Montgomery County.

Impervious surfaces such as roofs, driveways, parking lots, and streets prevent precipitation from entering the ground and the groundwater where it completes the hydrologic cycle. Instead stormwater is collected and either sent to a stormwater facility or discharged directly to the streams without control. Older, urbanized areas of the county without stormwater controls bear witness to the devastation visited upon nearby stream valleys, which were blasted by incredible volumes of water, sediment, and pollution, changing from gurgling, bucolic streams to 50 foot wide lifeless channels with toppled trees, exposed sewer lines, and deeply cut and eroded banks.

What is the Water Quality Protection Charge (WQPC)?

The WQPC will appear as a line item on property tax bills and will pay for the structural maintenance of stormwater management facilities. The Water Quality Protection Charge is the result of years of study, recommendations and hard work by citizens serving on work groups and task forces, County Council Staff and the Department of Environmental Protection.

Who will pay?

The charge will be paid by all residential property owners and any associated nonresidential property owners.

What are associated nonresidential properties?

An associated nonresidential property is any nonresidential property from which stormwater drains into a stormwater management facility that primarily serves one or more residential properties. Some examples of associated nonresidential properties could be: a restaurant that has a parking lot draining into a neighborhood stormwater pond, a church parking lot draining into a neighborhood pond, or a private school that has sidewalks, parking lots and outbuildings draining to a residential pond or other type of stormwater management structure.

How has the charge been determined?

The charge is based on the average amount of square feet of roof, sidewalk and driveway for a single-family dwelling. Wet weather cannot penetrate these "impervious" surfaces, thereby washing pollutants such as oil and grease from driveways, as well as fertilizers, pesticides, and

[continued, over]

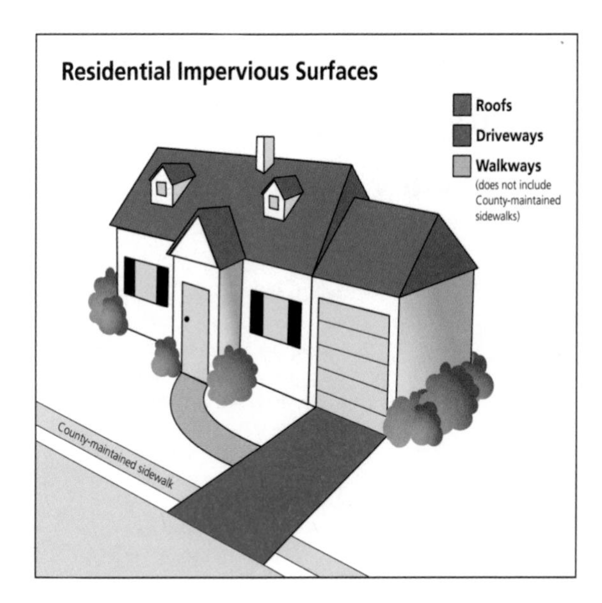

pet waste from yards and turf areas either into nearby streams or into a stormwater management structure. Accumulating stormwater also can erode stream banks if not properly managed by well-maintained ponds, sand filters, infiltration trenches or other stormwater management structures. The average impervious surface has been calculated to be 2,406 square feet and is the Equivalent Residential Unit (ERU) or the base unit for calculating the Water Quality Protection Charge.

Associated nonresidential structures are billed as multiples of the ERU. If a restaurant has 24,060 square feet of imperviousness, then the property owner will pay ten times the ERU.

Condominium and apartment charges will be calculated based on the amount of imperviousness and will be billed as multiples of the ERU. Townhomes will be billed at one-third of an ERU. Commercial and other land use classes that have on-site facilities that do not drain to residential facilities will not pay the charge but will be required to maintain their own structure.

How much is the WQPC?

The WQPC rate is determined by the costs of structural maintenance for residential and associated nonresidential stormwater facilities divided by the number of ERU's. Currently the intent of the law is to perform structural maintenance, although other program aspects, such as inspecting and repairing stormdrains, performing structural maintenance on nonresidential facilities not previously part of the program, and offering incentives through credits and exemptions, could be added to the program.

Additions to the program will require additional increases to the rate. As of March 2002, the proposed rate is $12.75 per ERU. The County Council will set the rate of the Water Quality Protection Charge on an annual basis. A public hearing will be held annually before the Council sets the rate.

What maintenance will be done by the County?

The County will perform structural maintenance on the stormwater facilities. Structural maintenance is defined by the law as: the inspection, construction, reconstruction, modification, or repair of any part of a storm water management facility undertaken to assure that the facility remains in the proper working condition to serve its intended purpose and prevent structural failure. Structural maintenance does not include landscaping, grass cutting, or trash removal.

What maintenance will remain the responsibility of the property owner?

The property owner will be responsible for the aesthetic maintenance around the facility, including trash removal and grass cutting. This routine maintenance is critical and if neglected can lead to expensive maintenance problems that will be the responsibility of the property owner.

Will the County own the property around the facility?

No, the property owner will be responsible for recording easements and covenants in the Montgomery County land records that allow the County to perform the maintenance on the facility.

When will the County accept stormwater facilities into the program?

The effective date of the legislation is March 1, 2002. The Department of Environmental Protection (DEP) will be accepting applications beginning on that date. The legislation specifies that any stormwater facilities granted final approval by the Department of Permitting Services up to that date are "existing" stormwater management structures and any facilities granted approval after that date are classified as "new" stormwater management structures.

Existing facilities will have to prepare new easement documents and new facilities will automatically become part of the program.

How will structures enter the County program?

The owner must make any structural repairs needed to place the facility in proper working condition, as determined by the Department of Environmental Protection usually to asbuilt standards if available, before the County enters into an agreement with the owner that obligates the County to

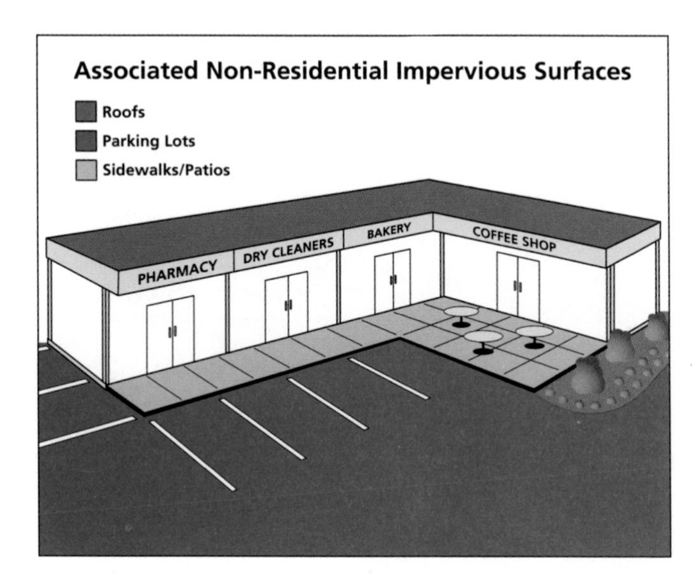

assume responsibility for structural maintenance of the facility.

After the owner and the County have agreed that the County will assume responsibility for structural maintenance, the owner must record the easement and any other agreements executed in conjunction with the easement that are binding on subsequent owners of land served by the facility in the County land records. The owner must deliver a certified copy of each recorded document to the Department of Environmental Protection. Structural maintenance becomes the responsibility of DEP after the documents are delivered to that Department.

Underground structures (oil/grit separators, underground storage structures, underground sand filters and water quality inlets) must have been cleaned and inspected within three months of final recordation date of easement documents. Above ground structures (dry ponds, wet ponds, and surface sand filters) must have been cleaned and inspected within twelve months of final recordation date of easement documents.

Is the Water Quality Protection Charge deductible from my Federal Income Tax?

No. The Water Quality Protection Charge is not deductible from Federal or State Income Taxes.

Why do I have to pay a WQPC?

The County is required by the Montgomery County Code to provide stormwater management facilities and services that control the quantity and quality of runoff entering the streams and rivers in the County, including the structural maintenance of those facilities. Developers generally pay for construction. Funding is not provided by federal or state governments for the maintenance of these facilities.

Why do I have to pay a WQPC for my restaurant but the restaurant two blocks away is not charged?

Your restaurant drains to a Stormwater management facility that primarily drains residential properties. The other restaurant drains to a Stormwater management facility that drains only non - residential properties.

Do tax exempt properties (if they are considered an associated nonresidential property) have to pay?

Yes, because it is a charge, not a property tax. Property taxes are based on the assessed value of the property. The WQPC is assessed based on how much the property contributes to the amount of Stormwater runoff from the property.

If tax exempt properties have to pay, why don't government-owned facilities (that are considered associated non-residential properties) - Federal, state, or local?

All government properties are not charged. An exemption is provided to State and Federal properties under Maryland state law. However, the County is making every effort to enter into Memoranda of Understanding with these entities to get them to agree to pay.

I own a farm. Why is my agricultural property (if it is an associated nonresidential property) being assessed this charge?

Farmhouses are being treated as residential properties. You are being charged based on the impervious area of your farmhouse, driveway, etc., not your whole property area. Cropland and pastureland are not charged.

Will revenues be spent throughout the County?

The Stormwater maintenance program is County-wide. However, the Cities of Rockville, Gaithersburg, and Takoma Park will not be included in this program because they are already implementing a Stormwater maintenance program in their own respective cities. Takoma Park will continue to assess its own stormwater fee to its residents.

[continued, over]

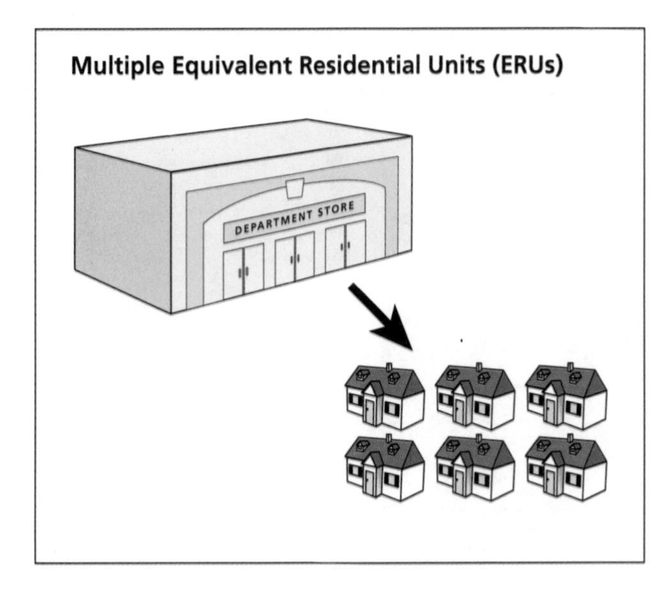

How was the amount of impervious surfaces determined?

Impervious surfaces were determined by analyzing a statistically significant number of residential parcels in the geographic information system (GIS) available from Montgomery County and the Maryland-National Capital Park and Planning Commission (M-NCPPC).

Do I have to pay for any undeveloped properties that I own?

No, because there is no impervious area associated with your property.

Do I have to pay for any unoccupied developed properties that I own?

Yes, because that property contains impervious area.

My neighborhood roads have drainage problems. Who do I call to get some action on these problems?

You can call the County Department of Public Works and Transportation at (240) 777-7623.

Why do I have to pay when I do not have any drainage problems?

Everyone in the County benefits from the stormwater maintenance program. If stormwater runs off your property, the County must have a program and funding to manage the increase in runoff and pollutants.

What happens if I don't pay or I pay late?

Interest on the overdue payment accrues according to the same schedule and at the same rate charged for delinquent real property taxes until the owner has remitted the outstanding payment and interest. An unpaid Charge is subject to all penalties and remedies that apply to unpaid real property taxes. If the unpaid Charge becomes a lien against the property, the lien has the same priority as a lien imposed for nonpayment of real property taxes.

Can I appeal the charge?

If a property owner believes that a Charge has been assigned or calculated incorrectly, the property owner may petition the Director for an adjustment by submitting a written request in a form acceptable to the Department of Finance within 21 days after the property owner receives a bill for the Charge. The request must state the grounds for the property owner's petition.

Why is the stormwater management program not funded by tax revenues?

It has been funded in the past by tax revenues. However, the WQPC is fairer than a stormwater tax based on the assessed value for the real property, because: The charge is based on each property's actual contribution to stormwater runoff. Each property contributes a fair and equitable share towards the overall cost of the stormwater maintenance program.

For more information:

Department of Environmental Protection / Montgomery County, Maryland
255 Rockville Pike, Suite 120, Rockville, MD 20850
240.777.7770 fax: 240.777.7765
e-mail: help@askDEP.com

Securing the Future

Investing in Our Infrastructure

Set on the eastern bank of the Connecticut River, the City of Springfield was established in 1636 and has since grown into a dynamic center of business and commerce. The city today supports a population of 156,000 within a greater region of half a million residents.

> *"The Springfield Water and Sewer Commission improves the quality of life for our community through public health protection, environmental stewardship and support of sustainable economic development."*
>
> *SWSC Vision Statement*

Springfield's diversity is represented by a blend of the historic and new. The City's vitality is reflected in investments in the community like the Civic Center expansion and the Basketball Hall of Fame. Historic structures have also been carefully restored to distinguish Springfield's place in national history.

Ongoing investments in the community contribute to the quality of life in Springfield. Maintaining this quality requires attention to the water and sewer systems that lie underground.

Underneath the streets of greater Springfield and Ludlow lies another historic artifact: the infrastructure that supports our water and sewer services. A network of hundreds of miles of water and sewer pipes extends underneath our streets. In combination with reservoirs, pumping stations, and water and wastewater treatment facilities, these pipelines bring clean water into every home and business—and they carry wastewater to the treatment plant so that we can protect the quality of our river by returning clean water to the environment.

How do infrastructure investments support our quality of life?

As utilities across the nation strain to meet budget pressures, the water and wastewater industry is taking a hard look at its business processes and making changes to follow private industry management practices. To that end, in 1996, the City established the SWSC to operate and maintain its water and sewer assets. SWSC has evolved into an organization run on sound business principles, as reflected by its formal mission statement:

> We will develop, operate and maintain our water and wastewater systems in a cost-effective manner by:
>
> - providing quality services to our customers,
> - protecting and maintaining an adequate, uninterrupted and high quality water supply,
> - providing effective drinking water treatment and distribution and supporting fire protection,
> - collecting and treating sewage to environmental standards beyond permit requirements, and returning it safely to the river,
> - developing and maintaining a safe, professional workforce, and
> - building alliances with communities and educating future generations about the importance of protecting our water resources.

Springfield's infrastructure dates back to the 1800s, and as in many other cities, it is aging and in need of repair and replacement. We now face a substantial financial challenge to keep our underground pipes and valves operating safely so that we can preserve the quality of life we enjoy above ground. In the City today, we experience an average of one water main break a week. Routinely responding to a growing number of emergencies is a costly—and risky—way to manage our infrastructure.

Springfield is not alone in this challenge. A recent report by the American Water Works Association (AWWA) on the nation's drinking water infrastructure found that spending money on pipe replacement must triple over the next 30 years to maintain safe, reliable drinking water systems. Putting off the investment only increases the cost.

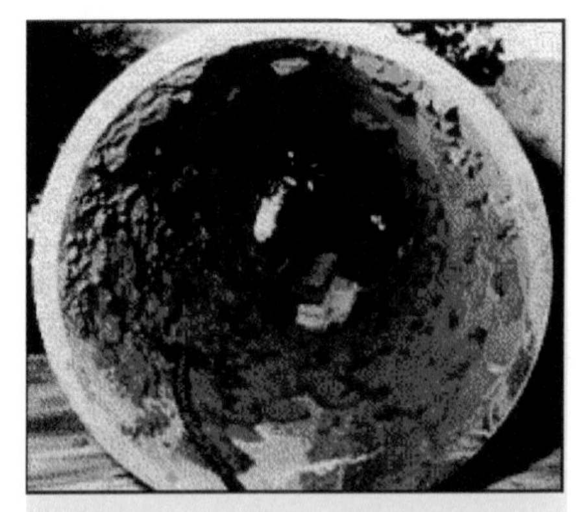

While progress continues above the ground, our underground systems have continued to age. 42% of our water pipelines are made of unlined cast-iron pipe—much of which is now clogged with rust and debris.

What have we done to improve the management of the system?

Water main break floods streets

Businesses try to stay afloat after water damages buildings

The Springfield Water and Sewer Commission knows that it must spend its ratepayers' money wisely on projects that will benefit its customers. Some examples of how SWSC is working hard to make every dollar count:

- Efficiencies gained by reorganization and a new public/private partnership at the wastewater plant have resulted in \$5 million in annual savings
- A planned meter replacement program will provide automatic, accurate readings without the inconvenience of meter readers entering properties
- A new financial system will improve accountability
- Monthly billing will provide customers with up-to-date information for budgeting purposes and for managing water use
- A newly implemented collection program provides accountability and fairness for all ratepayers
- Water conservation programs help preserve limited supplies and decrease individual bills.

Springfield's rates are far lower than those of most surrounding communities, and have not kept up with inflation or national trends

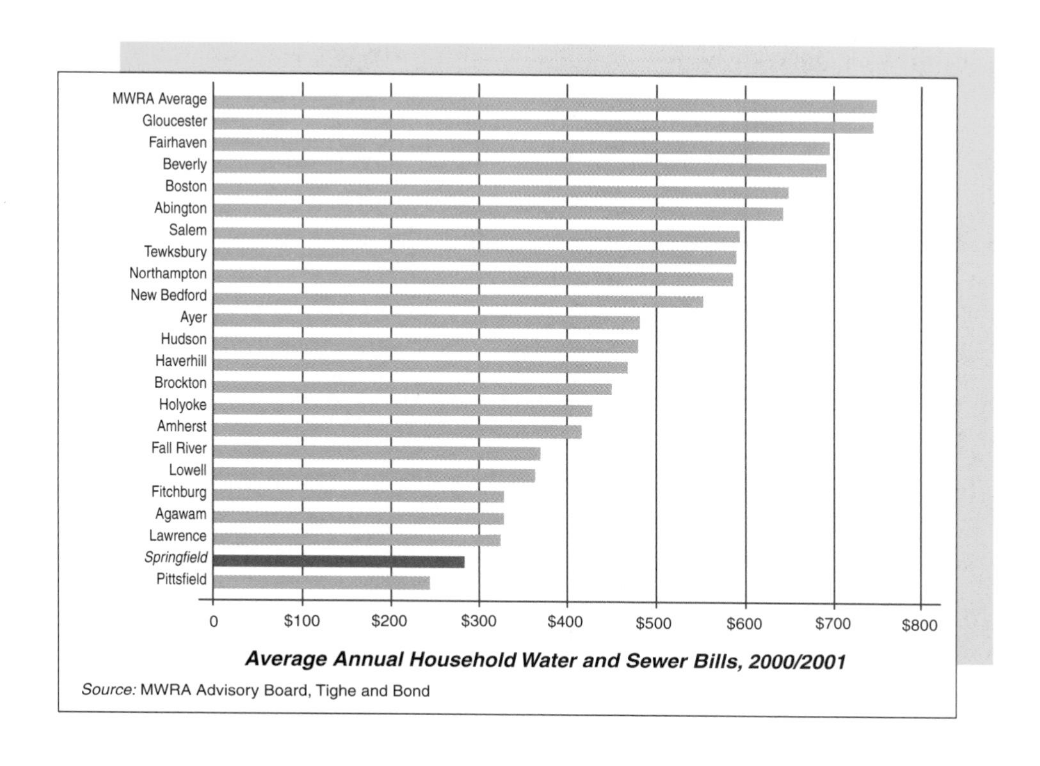

Average Annual Household Water and Sewer Bills, 2000/2001

Source: MWRA Advisory Board, Tighe and Bond

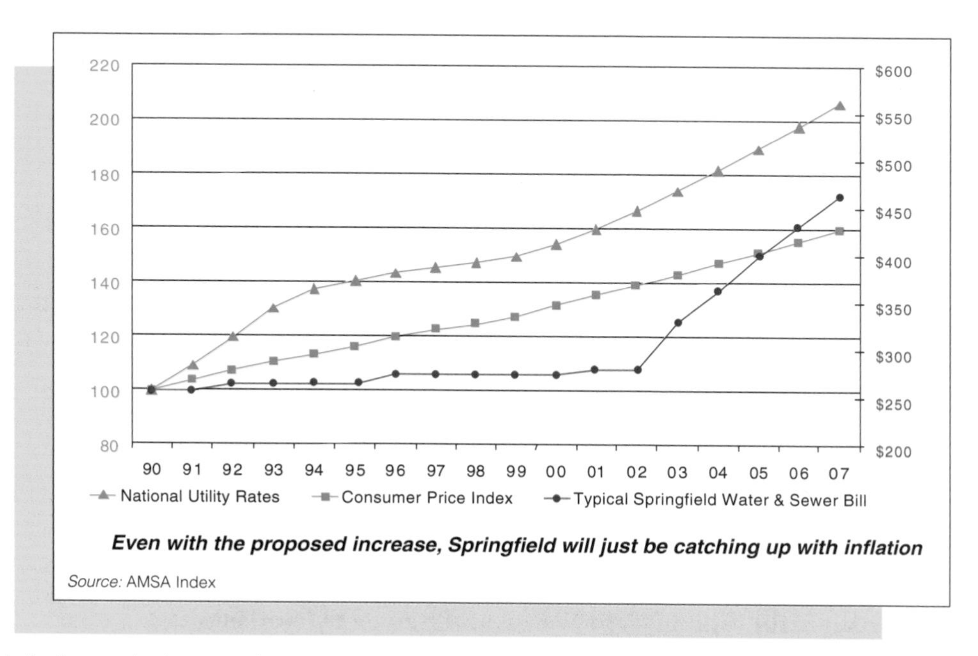

Even with the proposed increase, Springfield will just be catching up with inflation

Source: AMSA Index

The Commission is proposing incremental water and sewer rate increases to address the need for additional revenue and meet increased costs. The new rates represent an average annual increase of 11% over the five-year period. The typical household water and sewer bill is estimated to increase an average of $36.65 annually—that means an additional $3.05 per household per month. The rates will be restructured to eliminate the existing minimum water charge and administrative sewer charge, replacing it with a set water and sewer service charge to cover fixed administrative costs. The Commission will continue to offer a senior citizen allowance of $30 per year.

What investment is required to secure our future?

The Commission has developed a comprehensive, five-year capital improvement program to invest approximately $70 million in the water and sewer systems that serve our customers. Like most cities in the United States, Springfield has not historically budgeted for infrastructure renewal and replacement. More funding is needed to address a significant backlog of repair and replacement needs. Planned **renewal and replacement** is much less costly—and certainly less disruptive to residents and businesses—than emergency replacement. In the long run, we all benefit.

In addition, the SWSC is under a federal Administrative Order to reduce **combined sewer overflows**, which will force the spending of about $30 million over the next five years. In the wake of the tragic events of 9/11, heightened concern for the safety of our water supply means we will invest about $11 million in increased spending on **security** and **water treatment plant** improvements.

A water and sewer infrastructure replacement program will increase the reliability of our underground life support systems

The Springfield Water and Sewer Commission is planning a $25 million targeted program to fix pipes and upgrade facilities to support the City's economic development, environmental and public health goals. Some of the improvements planned are shown on the map below.

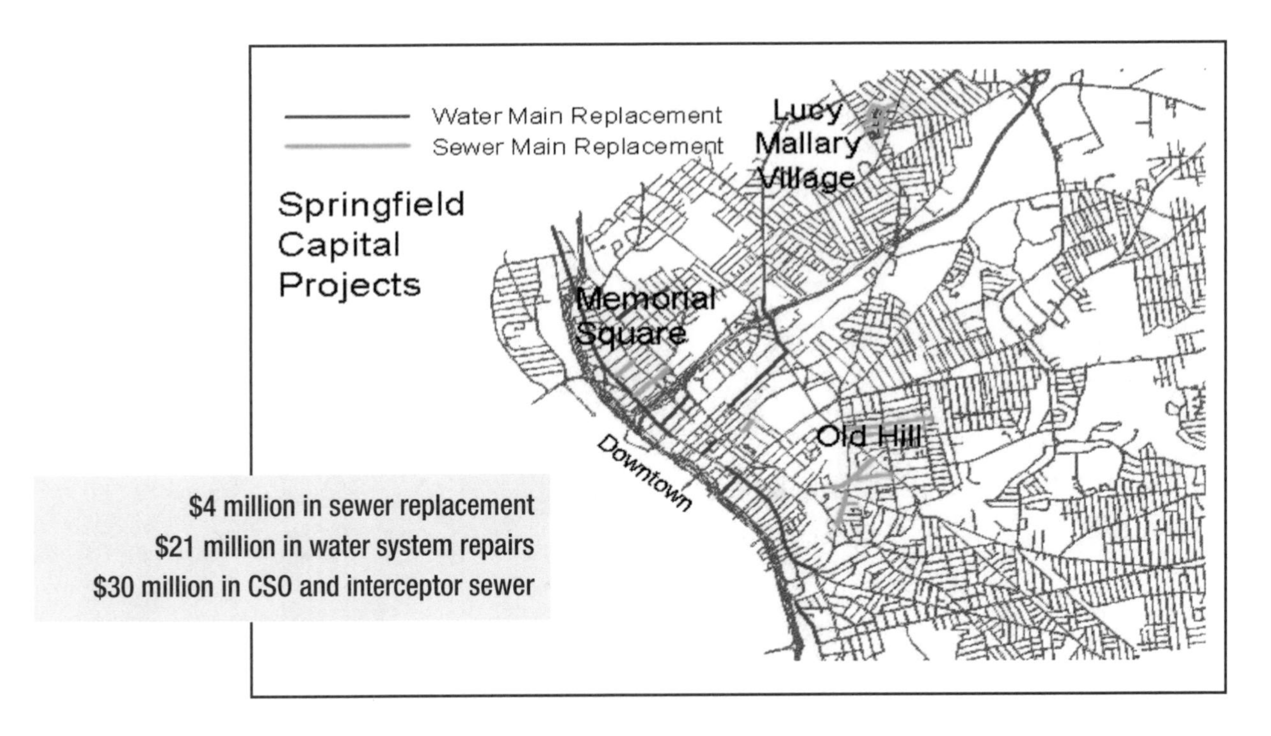

For more information, please contact Kathy Pedersen at 413-787-6256 x111, kathy.pedersen@waterandsewer.org or visit the Springfield Water and Sewer Commission web site at www.waterandsewer.org.

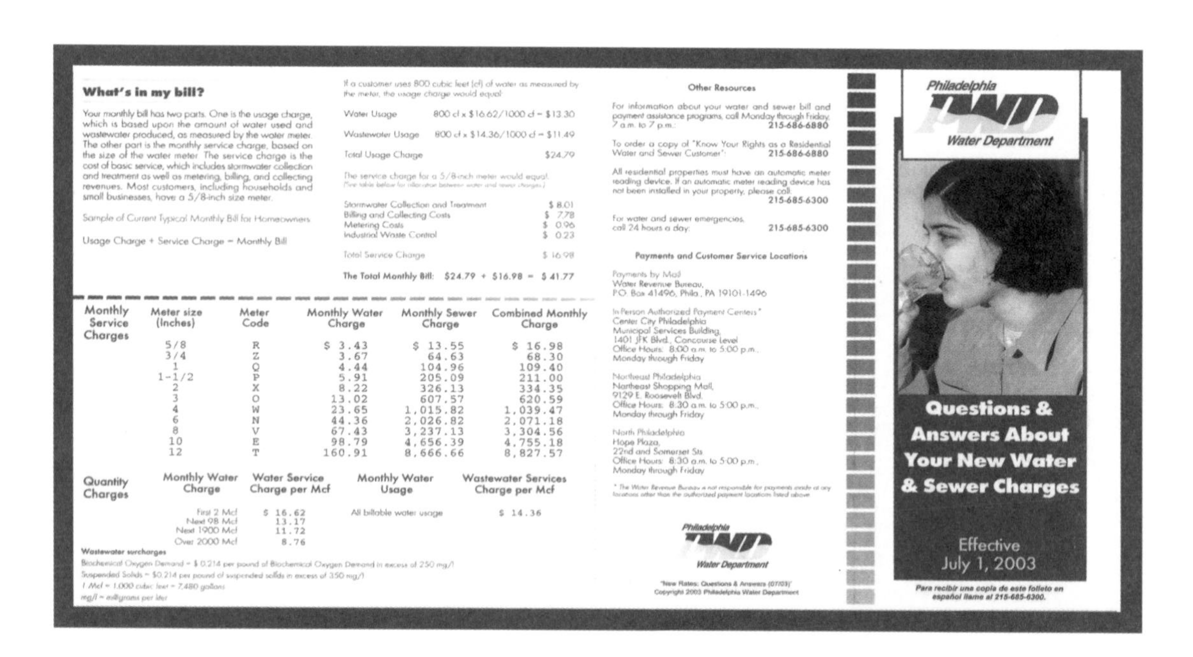

What's in my bill?

Your monthly bill has two parts. One is the usage charge, which is based upon the amount of water used and wastewater produced, as measured by the water meter. The other part is the monthly service charge, based on the size of the water meter. The service charge is the cost of basic service, which includes stormwater collection and treatment as well as metering, billing, and collecting revenues. Most customers, including households and small businesses, have a 5/8-inch size meter.

Sample of Current Typical Monthly Bill for Homeowners

Usage Charge + Service Charge = Monthly Bill

If a customer uses 800 cubic feet (cf) of water as measured by the meter, the usage charge would equal:

Water Usage	800 cf x $16.62/1000 cf =	$13.30
Wastewater Usage	800 cf x $14.36/1000 cf =	$11.49
Total Usage Charge		$24.79

The service charge for a 5/8-inch meter would equal:
(See table below for allocation between water and sewer charges.)

Stormwater Collection and Treatment	$ 8.01
Billing and Collecting Costs	$ 7.78
Metering Costs	$ 0.96
Industrial Waste Control	$ 0.23
Total Service Charge	$ 16.98

The Total Monthly Bill: $24.79 + $16.98 = $ 41.77

Monthly Service Charges	Meter size (Inches)	Meter Code	Monthly Water Charge	Monthly Sewer Charge	Combined Monthly Charge
	5/8	R	$ 3.43	$ 13.55	$ 16.98
	3/4	Z	3.67	64.63	68.30
	1	Q	4.44	104.96	109.40
	1-1/2	P	5.91	205.09	211.00
	2	X	8.22	326.13	334.35
	3	O	13.02	607.57	620.59
	4	W	23.65	1,015.82	1,039.47
	6	N	44.36	2,026.82	2,071.18
	8	V	67.43	3,237.13	3,304.56
	10	E	98.79	4,656.39	4,755.18
	12	T	160.91	8,666.66	8,827.57

Quantity Charges	Monthly Water Charge	Water Service Charge per Mcf	Monthly Water Usage	Wastewater Services Charge per Mcf
	First 2 Mcf	$ 16.62	All billable water usage	$ 14.36
	Next 98 Mcf	13.17		
	Next 1900 Mcf	11.72		
	Over 2000 Mcf	8.76		

Wastewater surcharges

Biochemical Oxygen Demand = $ 0.214 per pound of Biochemical Oxygen Demand in excess of 250 mg/l

Suspended Solids = $0.214 per pound of suspended solids in excess of 350 mg/l

1 Mcf = 1,000 cubic feet = 7,480 gallons

mg/l = milligrams per liter

Other Resources

For information about your water and sewer bill and payment assistance programs, call Monday through Friday, 7 a.m. to 7 p.m.: **215-686-6880**

To order a copy of "Know Your Rights as a Residential Water and Sewer Customer": **215-686-6880**

All residential properties must have an automatic meter reading device. If an automatic meter reading device has not been installed in your property, please call: **215-685-6300**

For water and sewer emergencies, call 24 hours a day: **215-685-6300**

Payments and Customer Service Locations

Payments by Mail
Water Revenue Bureau,
P.O. Box 41496, Phila., PA 19101-1496

In-Person Authorized Payment Centers*
Center City Philadelphia
Municipal Services Building,
1401 JFK Blvd., Concourse Level
Office Hours: 8:00 a.m. to 5:00 p.m.,
Monday through Friday

Northeast Philadelphia
Northeast Shopping Mall,
9129 E. Roosevelt Blvd.
Office Hours: 8:30 a.m. to 5:00 p.m.,
Monday through Friday

North Philadelphia
Hope Plaza,
22nd and Somerset Sts.
Office Hours: 8:30 a.m. to 5:00 p.m.,
Monday through Friday

* The Water Revenue Bureau is not responsible for payments made at any locations other than the authorized payment locations listed above.

Philadelphia Water Department

"New Rates: Questions & Answers (07/03)"
Copyright 2003 Philadelphia Water Department

Philadelphia Water Department

Questions & Answers About Your New Water & Sewer Charges

Effective July 1, 2003

Para recibir una copia de este folleto en español llame al 215-685-6300.

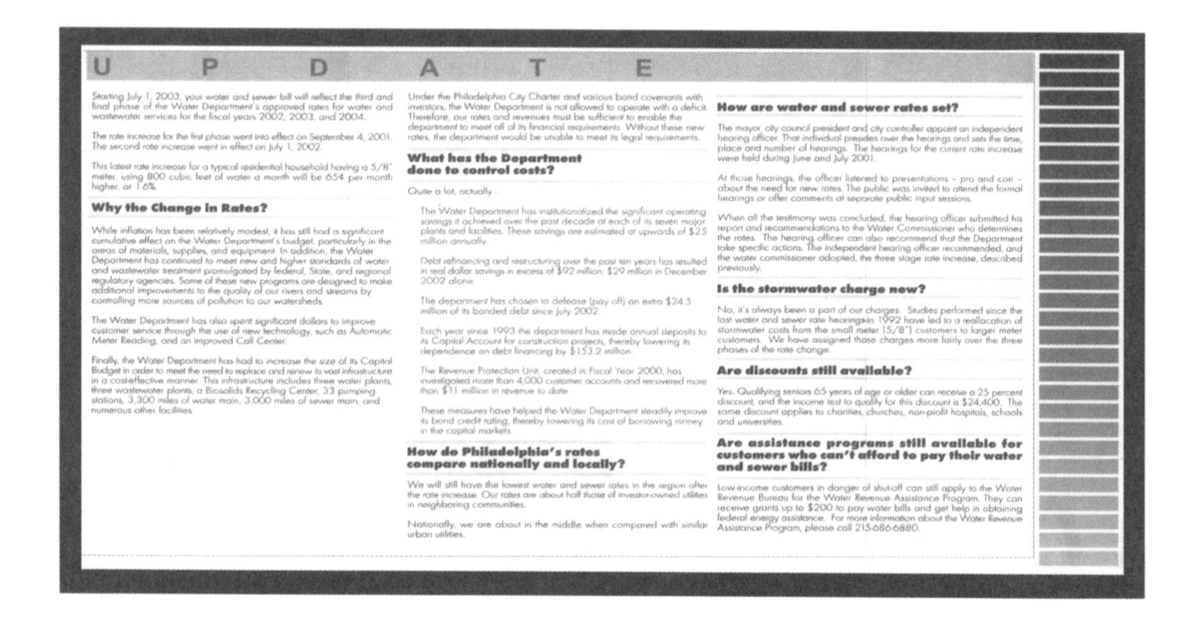

U P D A T E

Starting July 1, 2003, your water and sewer bill will reflect the third and final phase of the Water Department's approved rates for water and wastewater services for the fiscal years 2002, 2003, and 2004.

The rate increase for the first phase went into effect on September 4, 2001. The second rate increase went in effect on July 1, 2002.

This latest rate increase for a typical residential household having a 5/8" meter, using 800 cubic feet of water a month will be 65¢ per month higher, or 1.6%.

Why the Change in Rates?

While inflation has been relatively modest, it has still had a significant cumulative effect on the Water Department's budget, particularly in the areas of materials, supplies, and equipment. In addition, the Water Department has continued to meet new and higher standards of water and wastewater treatment promulgated by federal, State, and regional regulatory agencies. Some of these new programs are designed to make additional improvements to the quality of our rivers and streams by controlling more sources of pollution to our watersheds.

The Water Department has also spent significant dollars to improve customer service through the use of new technology, such as Automatic Meter Reading, and an improved Call Center.

Finally, the Water Department has had to increase the size of its Capital Budget in order to meet the need to replace and renew its vast infrastructure in a cost-effective manner. This infrastructure includes three water plants, three wastewater plants, a Biosolids Recycling Center, 33 pumping stations, 3,300 miles of water main, 3,000 miles of sewer main, and numerous other facilities.

Under the Philadelphia City Charter and various bond covenants with investors, the Water Department is not allowed to operate with a deficit. Therefore, our rates and revenues must be sufficient to enable the department to meet all of its financial requirements. Without these new rates, the department would be unable to meet its legal requirements.

What has the Department done to control costs?

Quite a lot, actually -

The Water Department has institutionalized the significant operating savings it achieved over the past decade at each of its seven major plants and facilities. These savings are estimated at upwards of $25 million annually.

Debt refinancing and restructuring over the past ten years has resulted in real dollar savings in excess of $92 million; $29 million in December 2002 alone.

The department has chosen to defease (pay off) an extra $24.5 million of its bonded debt since July 2002.

Each year since 1993 the department has made annual deposits to its Capital Account for construction projects, thereby lowering its dependence on debt financing by $153.2 million.

The Revenue Protection Unit, created in Fiscal Year 2000, has investigated more than 4,000 customer accounts and recovered more than $11 million in revenue to date.

These measures have helped the Water Department steadily improve its bond credit rating, thereby lowering its cost of borrowing money in the capital markets.

How do Philadelphia's rates compare nationally and locally?

We will still have the lowest water and sewer rates in the region after the rate increase. Our rates are about half those of investor-owned utilities in neighboring communities.

Nationally, we are about in the middle when compared with similar urban utilities.

How are water and sewer rates set?

The mayor, city council president and city controller appoint an independent hearing officer. That individual presides over the hearings and sets the time, place and number of hearings. The hearings for the current rate increase were held during June and July 2001.

At those hearings, the officer listened to presentations - pro and con - about the need for new rates. The public was invited to attend the formal hearings or offer comments at separate public input sessions.

When all the testimony was concluded, the hearing officer submitted his report and recommendations to the Water Commissioner who determines the rates. The hearing officer can also recommend that the Department take specific actions. The independent hearing officer recommended, and the water commissioner adopted, the three stage rate increase, described previously.

Is the stormwater charge new?

No, it's always been a part of our charges. Studies performed since the last water and sewer rate hearings in 1992 have led to a reallocation of stormwater costs from the small meter (5/8") customers to larger meter customers. We have assigned those charges more fairly over the three phases of the rate change.

Are discounts still available?

Yes. Qualifying seniors 65 years of age or older can receive a 25 percent discount, and the income test to qualify for this discount is $24,400. The same discount applies to charities, churches, non-profit hospitals, schools and universities.

Are assistance programs still available for customers who can't afford to pay their water and sewer bills?

Low-income customers in danger of shut-off can still apply to the Water Revenue Bureau for the Water Revenue Assistance Program. They can receive grants up to $200 to pay water bills and get help in obtaining federal energy assistance. For more information about the Water Revenue Assistance Program, please call 215-686-6880.

Every day, 24/7, 365 days a year, Las Virgenes Municipal Water District is treating wastewater from more than 85,000 local residents. Our service is transparent, safe, environmentally friendly and energy efficient — and that's the way it should be.

Costs to provide service, however, have nearly doubled in the last 10 years—with skyrocketing energy, environmental, security and insurance prices. Cost containment efforts are no longer enough, making it necessary to raise customer charges.

What are the changes?

Sewer fees will continue to be billed every other month, as part of your water bill. To keep fees as low as possible, the increases will be stepped in over time. On January 1, 2004, sewer charges will go up $2 per month; then $1 per month on July 1, 2004; and $1.50 per month, a full year later, in July 2005. Customers with exceptionally low water use (12 Units* or less in a billing period) will receive a 10% discount on their sewer charges. (*A Unit is a measure of water use equal to 748 gallons.)

It will still cost less than $1 per day to have all the wastewater from your household treated and disposed, even after the changes take effect.

FAST FACTS
about sewer service

- 15,958 sewer connections in the LVMWD service area
- 56 miles of trunk sewer lines operating 24/7
- 9 million gallons of wastewater safely treated each day
- 75% (or almost 7 million gallons per day) of treated wastewater recycled as irrigation water
- ZERO gallons of recycled water discharged into Malibu Creek April 15 to November 15
- 6,650 wet tons of solid waste removed from wastewater each month and beneficially composted
- Less than $1 per day for a household's sewer service

Sewer Service Charges

How your sewer service rates will be changing and why.

Las Virgenes Municipal Water District
4232 Las Virgenes Road, Calabasas, CA
(818) 251-2200 www.L-VMWD.com

Why does it cost so much to treat and dispose of sewage?

Heightened regulations, energy costs, permit fees, chemical purchases, and insurance coverage continue to escalate. Environmental costs for our district are greater than for most utilities, because our service location is within the Santa Monica Mountains. For instance:

- Wastewater treatment is regulated through dozens of permits, costing hundreds of thousands of dollars annually. Fees for state permits nearly doubled this year.
- Security measures post 9/11 mean more testing, reinforced barriers, strengthened lockdowns and 24-hour surveillance each and every day under escalated alerts.
- Modernized disinfection, for greater community and worker safety, has increased chemical costs of wastewater treatment nearly 50%.
- Workers compensation costs throughout California doubled this year, and other insurance pricing has nearly tripled.
- Electricity costs have almost doubled over 10 years. LVMWD's electrical bill for wastewater treatment exceeds $2.5 million annually.
- Until 1998, surplus recycled water could be released into Malibu Creek - a process now prohibited for 7 months each year. LVMWD spends more than a half-million dollars annually to comply with this requirement, with diversions to the City of Los Angeles, spray disposal on local lands, and incentives for customers to more fully irrigate landscapes. Construction of other alternatives have already cost more than $175,000, with an additional $3.5 to $5 million anticipated in the next 2 years.

Can't LVMWD just "tighten its belt" and absorb the increases?

We have. Despite a near doubling of sewage treatment costs in the past decade, customer charges have increased minimally. Potable water rates have *been reduced*.

LVMWD has consistently worked to be more efficient, absorbed costs, and put off the need to increase rates for as long as it could.

- 122 staff now provide all LVMWD potable water, wastewater, and customer services - down from 137 a decade ago when there were nearly 10% fewer customers.
- Reorganization and streamlining has cut management positions by 25%.
- In summertime, some staff starts work as early as 3 am, so high energy consumption tasks can be completed in off-peak hours, when electric rates are lower.
- Electric price increases have been absorbed even in the aftermath of the energy crisis.
- This year, nearly $5 million is being drawn from district reserves, to fund critical construction and maintenance programs.
- Lease of district facilities brings in nearly $300,000 annually, including the accommodation of phone and communications antennas on tanks and hilltop structures.
- Technology now provides on-going process monitoring reducing staff callouts and the associated overtime costs.

LVMWD understands the need to "stretch" a dollar. That's why we've taken the steps listed above to enhance efficiency. We also understand the demands you face, which is why the rate changes are applied across a protracted period of time. We know an extra dollar or two each month is not a large increase but we also want you to know we care and we're taking every possible measure to see that the region's wastewater is treated in a safe, dependable, environmentally-sensitive manner.

The Good News

PWD Efforts to Control Costs and Improve Services

Following its last rate increase, implemented in 1993 to 1995, the PWD committed itself to optimize its operating costs and to maintain its rates into the new century. This we have done. However, despite our best efforts to stretch those rates even further, after six years of rate stability, we have notified City Council of our need to increase water and sewer rates, beginning in July 2001, to meet an anticipated shortfall of $134.8 million in revenues over our fiscal years 2002 -2004.

PWD typical residential customers, using approximately 800 cubic feet of water per month, will experience an average annual increase of 3.1 percent, over a three-year period. The typical residential customer will see an average annual increase of $1.23 per month in their water and sewer bill.

The PWD has implemented many new programs and strategies since its last rate increase in 1995 in response to customer comments gleaned during the last public hearing process, and in reaction to our own revitalized commitment to our mission. The PWD serves the Philadelphia region by providing integrated water, wastewater and stormwater services. The PWD operates three drinking water plants which treat an average of 280 million gallons of Delaware and Schuylkill river water each day, three wastewater plants which clean over 450 million gallons per day of sewage, a 73-acre biosolids recycling facility, a sophisticated testing laboratory to assure water quality, and a host of technical and administrative support services. In addition, the PWD maintains 3,300 miles of water mains, 3,000 miles of sewers, 75,000 stormwater drains, over 27,500 fire hydrants, and an extensive related infrastructure.

Primarily, we plan for, operate and maintain both the infrastructure and the organization necessary to provide healthy and high quality drinking water, to provide an adequate and reliable water supply for all household, commercial and community needs, and to sustain and enhance our region's watersheds and the quality of life within our watersheds by managing wastewater and stormwater effectively. To fulfill this mission, we also seek to be customer-focused, delivering services in a fair, equitable, and cost-effective manner, with a pledge to public involvement.

A rate increase, although certainly not the most dispassionate topic, encourages customer communication in a vital way. We would like to take this opportunity to outline how we have met our commitments to our customers, following the last round of public hearings, and to share some of the good news and

---

ve Water/$ave Dollars

month this year, we'll be presenting a simple
seful tip for reducing your water use and
ing your water bill. Most people don't realize
much water can go down the drain from a
ng faucet, or that more than half the water
onsume is used outside the house to irrigate
. Knowing some simple ways to save water can
ou big on your water bill each month. Check
n Water's web site at www.cityoftucson.org/water/
ore water saving tips.

ruary Tip:

Lose Your Leaks

Your WATER Connection

News & Tips for Tucson Water Customers

Water 101

Throughout the year, Tucson Water receives hundreds of questions
customers about water
s ranging from water quality
stem and infrastructure
tions and everything in
een. Our customers are very
informed and inquisitive
t water. Perhaps it's because
understand so well how
rtant water is to a desert
munity like Tucson. Because
s high level of interest, we
eginning a new occasional
mn called Water 101 that will
in an aspect of your water
m. ***If you have a question you'd***
o have answered here, or if you
a suggestion for a topic, call us
1-4331 or email to
Web1@ci.tucson.az.us. We
you find Water 101
mative.

03 http://www.cityoftucson.org/water/

Save Water/$ave Dollars

March Water Conservation Tip:

Irrigation Timer Rx for Healthy Plants & Wise Water Use

Spring is a good time to check your outdoor watering practices to make sure your yard plants remain healthy during the hot summer season. This will also help you use water wisely outside – where we use the most water, especially during the summer.

1. How much water is enough? – Check the efficiency of your irrigation system by using a soil probe. Soil probes can be purchased at many garden centers, or you can use a long metal rod or even a long screwdriver. Push the probe into the soil after an irrigation cycle. It will stop when it hits dry soil, giving you an idea of how deep your watering has penetrated. Over time, you will also learn how fast the soil dries out.

2. Adjust your irrigation timer – As the days become warmer, landscape plants will require more frequent watering due to the soil drying out quicker. Remember to keep the duration the same, just alter the frequency of the watering. Monitor the soil moisture depth and the length of time it takes to dry out with your probe. Adjust your irrigation timer slowly adjusting the water frequency as the days get longer and hotter.

3. Check your irrigation system for leaks – Manually run your system to look for broken irrigation lines, missing emitters, or broken or misaligned sprinkler heads. Replace dead plants that are still being watered, or remove the emitter and plug the line to avoid this waste of water. Check the number of emitters on each plant to help avoid over watering.

Need more help? The Tucson Water Zanjero Program can help you manage your water use at your home or business. Call 791-3242 to schedule an appointment to have a trained Water Conservation Specialist check your indoor and outdoor water use.

A Reminder ...

Tucson Water's Automated Bill Payment Service is Available to Save You Time and Money

Make your water, sewer, or city garbage pickup payments automatically every month with Tucson Water's Automated Bill Payment Service. All you need is a valid checking or savings account, complete the application, and your payments will be automatically deducted each month from your account. You will still receive a statement from us and you'll save time and money every month. For more information and an application, visit our website at **www.cityoftucson.org/water** and click on "customer svcs" or call 791-3242.

March 2003 http://www.cityoftucson.org/water/

FAST FACTS ABOUT PWD'S RATE INCREASE PROPOSAL

How Much?

The PWD requires an increase in its water and sewer rates over a three-year period in the amount of 3.7 percent, 3.5 percent and 2.2 percent respectively for FY 2002, 2003 and 2004, to meet an anticipated shortfall of $134.8 million in revenue obligations. This will result in an average increase of $1.23 per month for typical residential customers using approximately 800 cubic feet of water per month.

When?

The PWD notified City Council on February 20, 2001. The PWD submitted its rate increase request with the Department of Records on March 23, 2001. The PWD plans to implement the first phase of its requested increase on July 1, 2001, the first day of its new fiscal year, following its public hearing process.

Why?

The proposed rate increase is required to allow the PWD to meet its operating expenses and revenue requirements associated with: debt service, inflation, wage increases—factors that have affected utilities nationwide. By comparison, the average rate of increase in the water utility industry in the U.S. has been approximately 12.7 percent per year between 1995 and 1999.

Last Increase?

The PWD increased its rates over a three year period between July 1, 1993 and July 1, 1995 by the following increments: 7 percent, 2.3 percent and 3.1 percent, the final two years based upon the Philadelphia Consumer Price Index (CPI).

For More Information

A number of formal and informal public hearings have been scheduled throughout the city to provide our customers with an opportunity to comment and ask questions. A schedule of these hearings will be available by calling 215.685.6300 and on the PWD's website at http://www.phila.gov/departments/water.

FACT SHEET

What Does the Water and Sewer Bill Include?

The monthly water and sewer bill contains two parts. One is the usage charge, which is the amount of water used and wastewater produced by a household or business as measured by the water meter. The other component is the service charge, a monthly charge based on the size of the water meter. The service charge is the cost of basic services which are provided to fund pipe maintenance, metering, billing and collections, and treatment associated with stormwater. Most PWD customers, including households and small businesses, have a 5/8-inch size meter.

Sample of Current Typical Monthly Bill for Homeowners:

Usage Charge + Service Charge = Monthly Bill

If a customer uses 800 cubic feet of water (cf) as measured by the meter, the usage charge would equal:

Water usage – 800 cf × $12.17/1000 cf = $9.74

Wastewater usage – 800 cf × $12.49/1000 cf = $9.99

Total Usage Charge = $19.73

The service charge for a 5/8-inch meter would equal:

Metering/Billing/Collecting	$6.15
Pipe Maintenance	$1.53
Industrial Waste Control*	$0.06
Stormwater Collection and Treatment**	$11.00
Total Service Charge =	$18.74

The Total Monthly Bill: $19.73 + $18.74 = $38.47

A total increase of 9.4 percent (implemented over a three year period beginning July 1, 2001) in this monthly bill will result in an average annual increase of $1.23 to the residential monthly bill.

---

* The wastewater produced by businesses and industries can contain the chemicals used in their manufacturing or production processes. That is why they are required to treat their wastewater "on-site" prior to emptying it into the city's sewer system. The PWD's Industrial Waste Unit inspects and enforces this pretreatment program.

** The PWD is responsible for the collection, drainage and treatment of stormwater to alleviate potential flooding and to keep stormwater free of pollutants to the best of its ability.

FACT SHEET

How Do PWD Rates Compare with Local and National Water and Wastewater Utilities?

The PWD currently provides services for the lowest cost in the region. The PWD's water rates are less than half those charged by most neighboring investor-owned utilities. Our rates will continue to be the most economical in the region despite our request for increased rates.

2000 Regional Residential* Water and Sewer Charges		
	Monthly Water Bill	Monthly Sewer Bill
Pennsylvania American Water†	$40.05	N/A
Philadelphia Suburban Water†	$35.72	N/A
New Jersey American Water†	$31.39	N/A
North Wales Water Authority†	$26.44	N/A
North Penn Water Authority†	$26.45	N/A
Doylestown Township	$20.33	$36.67
CCMUA (Camden County)‡	N/A	$26.25
Trenton	$18.12	$20.31
Philadelphia Water Department (current)	***$12.82***	***$15.46***
Philadelphia Water Department (proposed)	**$14.14**	**$15.87**

Source: Philadelphia Water Department

Rates in effect on November 22, 2000. Stormwater charges are excluded from sewer calculations, because many jurisdictions fund such services from the general tax base or a separate utility assessment. PWD's stormwater charges will **decrease** for residential customers in the proposed rates for FY 2002–FY 2004.

*Calculations based on 6230 gallons/month (833 cu.ft.).
†Sewer-only utility.
‡Water-only utilities.

On a national level, the PWD falls somewhere in the middle when compared to water utilities located in similar urban locations, although such comparisons are a bit more difficult due to vast differences in age and expanse of infrastructure, facilities and the availability of a consistent water supply. The PWD has a declining block rate structure for large quantity users (versus an inclining block structure) as it has the abundant resources of the Schuylkill and Delaware rivers to draw upon. Declining block rates are used for large metered customers who need large quantities of water as a part of their manufacturing or production operations. The volume charge for water per thousand cubic feet incrementally decreases as the meter size and volume of water used increases.

FACT SHEET

What is the Rate Stabilization Fund?

The Rate Stabilization Fund was created with the sale of the PWD's Series 1993 Revenue Bonds as a revenue resource to offset deficits in operating expenses as a means to temper the need for rate increases. It essentially provides a reserve fund for the PWD to draw upon, as a homeowner would do with a savings account, to fund expenses above and beyond those originally budgeted. The creation of this fund, mandated by the terms of our 1993 bond series, is a standard good financial practice. The fund assists in controlling rates in a fluctuating economy. The PWD's Capital Account, also a mandated fund, has been used to pay for the replacement of infrastructure in a "preventive" maintenance mode at an increased rate to accommodate our city's aging water and sewer mains. This fund enables the PWD to keep customer rate increases at a more manageable level when they are finally necessary, and lessens the department's need to sell bonds for its Capital Program.

Between Fiscal Years 1994 and 1998, approximately $100 million was deposited into the PWD's Rate Stabilization Fund as the PWD was able to meet its revenue obligations with some extra income to spare. However, since those years, the PWD has been drawing on this account to ensure that it continues to meet its bond covenants and other expenses. The PWD will continue to draw upon this fund into the future to alleviate the impact of a rate increase on our customers. Use of the Rate Stabilization Fund allowed PWD to avoid rate increases in FY 1999, 2000 and 2001. It will continue to be used in FY 2002, 2003 and 2004 to augment the moderate rate increase requested. Without this fund, the department's request and need for additional rates would be much higher.

FACT SHEET

PWD Assistance Programs

The PWD and WRB have long been leaders in providing assistance to customers who find it difficult to afford their utility bills. The special discounts and programs provided in FY00 are summarized in the chart below, and represent a major commitment to helping customers who might otherwise face service shutoffs. These programs provide a safety net for customers who require help in meeting their utility obligations.

Water Department and Water Revenue Bureau Assistance Programs			
Senior Citizen Discount	A 25 percent discount is provided for senior citizens 65 years of age or older, with a total household income <= $20,900/ year.	Administered by the Water Revenue Bureau.	Provided discounts to 47,000 seniors in FY00, at a total cost of $4.2 million.
Charitable Organization Discount	A 25 percent discount is provided for charities, churches, nonprofit hospitals, schools, and universities.	Administered by the Water Department and Water Revenue Bureau.	In FY00, the general charitable discount was used by 3,000 organizations at a total cost of $4.7 million.
Water Revenue Assistance Program (WRAP)	Grants of up to $200 on water bills are available to prevent shutoff for low-income customers (at or below 150 percent of poverty level). Assists customers in obtaining federal energy assistance.	Administered by the Water Revenue Bureau.	In FY00, provided City grants to 1,573 customers at a total cost of over $280,000.
Utility Emergency Services Fund (UESF)	Grant program to prevent shutoff for low-income customers (at or below 150 percent of poverty level). Provides up to $500 every other year ($250 UESF grant plus $250 matching Water Department credit).	Administered by the nonprofit UESF; with application help available from the Water Revenue Bureau.	Served 935 customers in FY00, requiring Water Department matching credits and administrative costs of over $350,000
Homeowners' Emergency Loan Program (HELP)	No-interest repair loan program for homeowners in imminent danger of shutoff because of a violation notice.	Administered by the Water Department.	Provided loans to 778 homeowners in FY00 at a total cost of over $1.7 million.
Conservation Assistance Program (CAP)	Provides water conservation devices and education to low-income customers (at or below 150 percent of poverty level), yielding average water usage savings of more than 25 percent for participants.	Administered by Neighborhood Energy Centers under a Water Department grant.	Served 1,773 households in FY00 at a total cost to the City of just under $400,000.
TOTAL:	The total FY00 cost of Water Department and Water Revenue Bureau assistance programs exceeded $11 million.		

CALENDAR FOR PHILADELPHIA WATER DEPARTMENT RATE PROCEEDINGS

Date	Action
March 19	Notice of Rate Change Filing with City Council
	Legal Advertising one day per week for three consecutive weeks—stating the an estimate of the proposed increase on average residential customer and that filing is available for inspection (Free Library—Regional Offices; Water Department; Records Department)
April 19	Regulations Filed with Department of Records Public Advocate Appointed
	Independent Hearing Officer Appointed
April 30	Pre-Hearing Conference Scheduled
May 1–28	Discovery Period (Formal and Informal)
May 28–June 1	Public Input Hearings (4 Hearings)
May 28	Technical Hearings (Initial hearing)
June 8	Technical Hearings (Close of record)
June 17	Simultaneous Initial Briefs (14 days after close of record)
June 24	Simultaneous Reply Briefs (7 days after initial briefs)
July 24	Report of Hearing Examiner (30 days after reply briefs filed)
August 3	Exceptions (10 days after Report of Hearing Examiner)
August 21	Decision of Water Commissioner (30 days after Hearing Examiner Report)
September 1	New Rates Effective (10 days after revised Regulations filed with Department of Records)

THE PHILADELPHIA WATER DEPARTMENT'S PROPOSED RATE INCREASE

Deputy Water Commissioner Bernard Brunwasser announced today that the Philadelphia Water Department is seeking an increase in its water and sewer rates beginning July 2001. The proposed new rates will be spread over a three-year period and are expected to result in average annual increases of $1.23 per month, or 3.1%, for a typical residential customer using 800 cubic feet of water. A typical senior citizen household can expect to see even smaller increases, on the order of 1.9%. This marks the first time the Water Department is seeking a rate increase in six years.

The proposed rate increases are needed to allow the Philadelphia Water Department to continue to meet all of its operating expenses and revenue requirements associated with debt service, inflation, new wage agreements, and new federal and State regulatory requirements.

Nationwide, rates for water and wastewater utilities have been increasing at average annual levels of more than five percent throughout the 1990s. "The Philadelphia Water Department has been able to hold the line on rates for the past six years by saving many millions of dollars in its wastewater treatment and bio-solids operations, and by refinancing more than $1.3 billion of its outstanding debt at much lower interest rates," said Deputy Commissioner Brunwasser. "The Water Department is much stronger financially and operationally since it last required rate relief. The employees and management of the department are proud of what they have accomplished and we are pleased that the rate increases we are proposing today are moderate and assure continued stability for our customers for at least the next three years."

The Philadelphia Water Department formally notified City Council on February xx, 2001, and will file its rate increase request with the Department Records on March xx, 2001. Plans to implement the first phase of these proposed rates are scheduled for July 1, 2001, the first day of the Water Department's new fiscal year, following the public hearing process.

A number of formal and informal public hearings will be scheduled throughout the city to provide our customers with an opportunity to comment and ask questions. A schedule of these hearings will be available in a few weeks by calling 215 685-6300, and on the Philadelphia Water Department's website at http://www.phila.gov/departments/water.

The Good News

PWD Efforts to Control Costs and Improve Services

Following its last rate increase, implemented in 1993 to 1995, the PWD committed itself to optimize its operating costs and to maintain its rates into the new century. This we have done. However, despite our best efforts to stretch these rates even further, after six years of rate stability, we have notified City Council of our need to increase water and sewer rates, beginning in July 2001, to meet an anticipated shortfall of $134.8 million in revenues over our fiscal years 2002 -2004.

PWD typical residential customers, using approximately 800 cubic feet of water per month, will experience an average annual increase of 3.1 percent, over a three-year period. The typical residential customer will see an average annual increase of $1.23 per month in their water and sewer bill.

The PWD has implemented many new programs and strategies since its last rate increase in 1995 in response to customer comments gleaned during the last public hearing process, and in reaction to our own revitalized commitment to our mission. The PWD serves the Philadelphia region by providing integrated water, wastewater and stormwater services. The PWD operates three drinking water plants which treat an average of 280 million gallons of Delaware and Schuylkill river water each day, three wastewater plants which clean over 450 million gallons per day of sewage, a 73-acre biosolids recycling facility, a sophisticated testing laboratory to assure water quality, and a host of technical and administrative support services. In addition, the PWD maintains 3,300 miles of water mains, 3,000 miles of sewers, 75,000 stormwater drains, over 27,500 fire hydrants, and an extensive related infrastructure.

Primarily, we plan for, operate and maintain both the infrastructure and the organization necessary to provide healthy and high quality drinking water, to provide an adequate and reliable water supply for all household, commercial and community needs, and to sustain and enhance our region's watersheds and the quality of life within our watersheds by managing wastewater and stormwater effectively. To fulfill this mission, we also seek to be customer-focused, delivering services in a fair, equitable, and cost-effective manner, with a pledge to public involvement.

A rate increase, although certainly not the most dispassionate topic, encourages customer communication in a vital way. We would like to take this opportunity to outline how we have met our commitments to our customers, following the last round of public hearings, and to share some of the good news and management goals that we have realized as a result of a myriad of initiatives.

Fiscal and Asset Management

PWD provides water and sewer services with the lowest rates in the region. This will hold true following the implementation of this rate request, continuing our trend of charging half the amount billed by neighboring utilities. How do we do this?

- Aggressive cost cutting strategies at our Biosolids Recycling Center (BRC) and our three wastewater plants have saved the department millions of dollars each year. The BRC has reduced its spending from $31.6 million in Fiscal Year (FY) 1993 to $17.6 million in FY 2000. Our three treatment plants have simultaneously reduced their cost of operation from $41.2 million to $31.4 million over the same period.

- The PWD has continued to gain significant cost savings in its energy bills through the successful implementation of a number of initiatives, including off-peak operation, conservation, and the installation of cogeneration and standby electric generating capacity at our Northeast and Southwest wastewater treatment plants. PWD saved approximately $1.2 million per year from FY 1994 through FY 1999 as a result of these initiatives. The Department has also saved more than $1 million annually as a result of the Citywide rate reduction negotiated by the Municipal Energy Office in 1996 and expects that lighting upgrades at its facilities will result in additional savings of $123,000 beginning in FY 2001.

FACT SHEET

Major Causes of PWD Rate Increase

The PWD was successful in keeping its rates stable between 1996 and 2000. Most of this success was due to unprecedented reductions in our operating costs and the refinancing of our revenue bonds. Unfortunately, during the same period, Philadelphia continued to lose a significant portion of its population and businesses, eroding the Water Department's rate base. Water and sewer services are also the most capital-intensive of utilities. As population shrinks, the fixed costs for maintaining and operating our water and wastewater treatment facilities and the 6,300 mile water main and sewer infrastructure, is spread over fewer customers.

Since the PWD last requested an increase in 1992, our total operating budget has grown to $442.7 million for FY 2002, an increase of $78 million. This $78 million difference, which represents a 21.4 percent increase over the nine-year period includes:

- An average annual increase of 2.4 percent
- A 27 percent increase in salaries and a 45.3 percent increase in fringe benefits
- Annual debt service is up $19.5 million, a 14.1 percent increase since FY 1993
- The PWD's interfund payment to the General Fund has increased by 16.9 percent, to a FY 2000 total of $9.8 million
- Inflation has increased the cost of materials (chemicals, equipment) by $5.8 million, or 23.1 percent
- The PWD is required to make an annual deposit of approximately $16 million from its Revenue Account to its Capital Account (this is a good thing, as it enables the PWD to fund its capital program, e.g., water main and sewer replacement, without requiring additional borrowing)
- The increased costs of state and federal drinking water and pollution prevention regulatory programs, e.g., Interim Enhanced Surface Water Treatment Rule, the Combined Sewer Overflow program and source water assessments

The proposed rate increase is designed to meet a project revenue shortfall of approximately $134.8 million over the period of FY 2002–FY 2004.

Securing the Future—
Investing in our Infrastructure

Springfield Water and
Sewer Commission
Clean Water – the Gift We Take for Granted
Our Vision: We make a difference
The Springfield Water and
Sewer Commission improves
the quality of life for our
community through...
■ public health protection
■ environmental stewardship, and
■ support of sustainable economic development
Working together, we can
build a better future

For more information, please contact:
Kathy Pederson, 413-787-6256 x111
kathypedersen@waterandsewer.org

Securing the Future—Investing in our Infrastructure

Springfield Water and Sewer Commission

Clean Water—the Gift We Take for Granted

Our Vision: We make a difference

The Springfield Water and Sewer Commission improves the quality of life for our community through...

- public health protection
- environmental stewardship, and
- support of sustainable economic development

Our Mission

The Springfield Water and Sewer Commission will develop, operate and maintain our water and wastewater systems in a cost-effective manner by:

- providing quality services to our customers,
- protecting and maintaining an adequate, uninterrupted and high quality water supply,
- providing effective drinking water treatment and distribution and supporting fire protection,
- collecting and treating sewage to environmental standards beyond permit requirements, and returning it safely to the river,
- developing and maintaining a safe, professional workforce, and
- building alliances with communities and educating future generations about the importance of protecting our water resources.

Historic under-investment in water and wastewater infrastructure is a documented nationwide concern

*"Pipes are expensive, but invisible.
Pipes are hearty, but ultimately mortal. ...
Increased expenditures are needed ...
The bills are now coming due, and
they loom large."*

*American Water Works Association
May 2001*

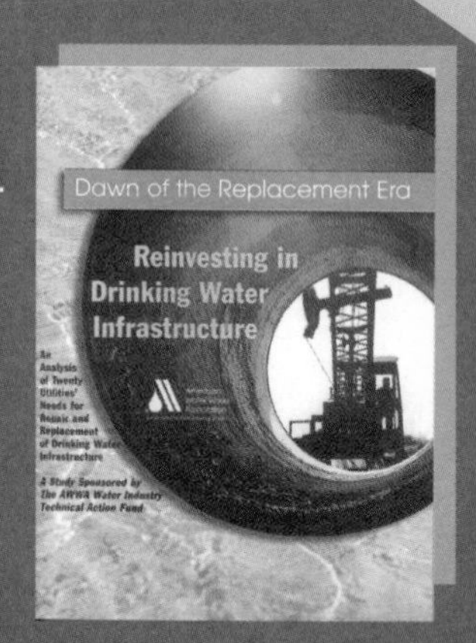

Springfield population trends are echoed by infrastructure investment

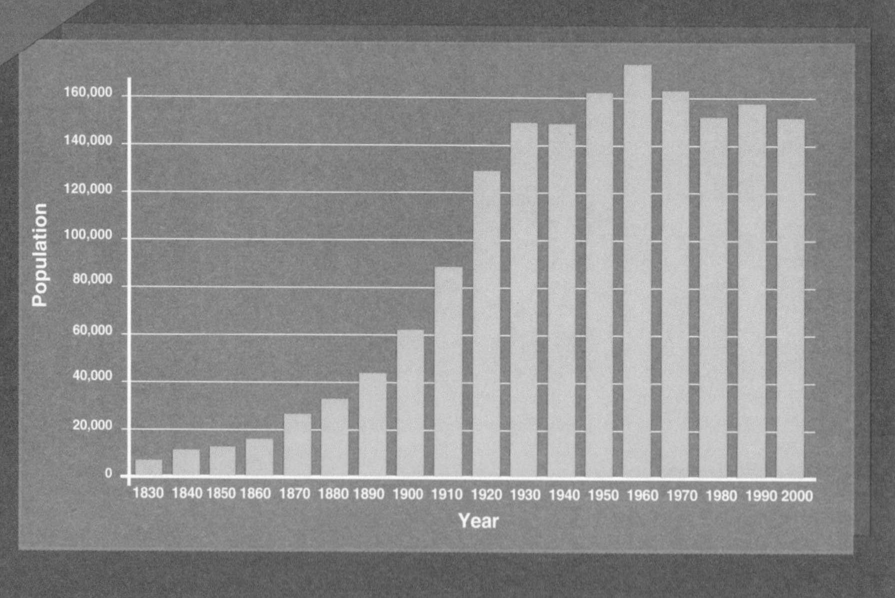

We don't want Springfield's pipes to look like this

Springfield's system includes 1,378,633 feet of unlined cast iron pipe—42% of the water distributiion system

Failing infrastructure is bad news for Springfield's economy

Water main break floods streets

Businesses try to stay afloat after water damages buildings

Sound water and wastewater systems are essential to economic development

"An uninterrupted, high quality supply of water and well-maintained infrastructure is critical to the economy of Greater Springfield."

—Tom McColgan
Springfield Economic Development Department

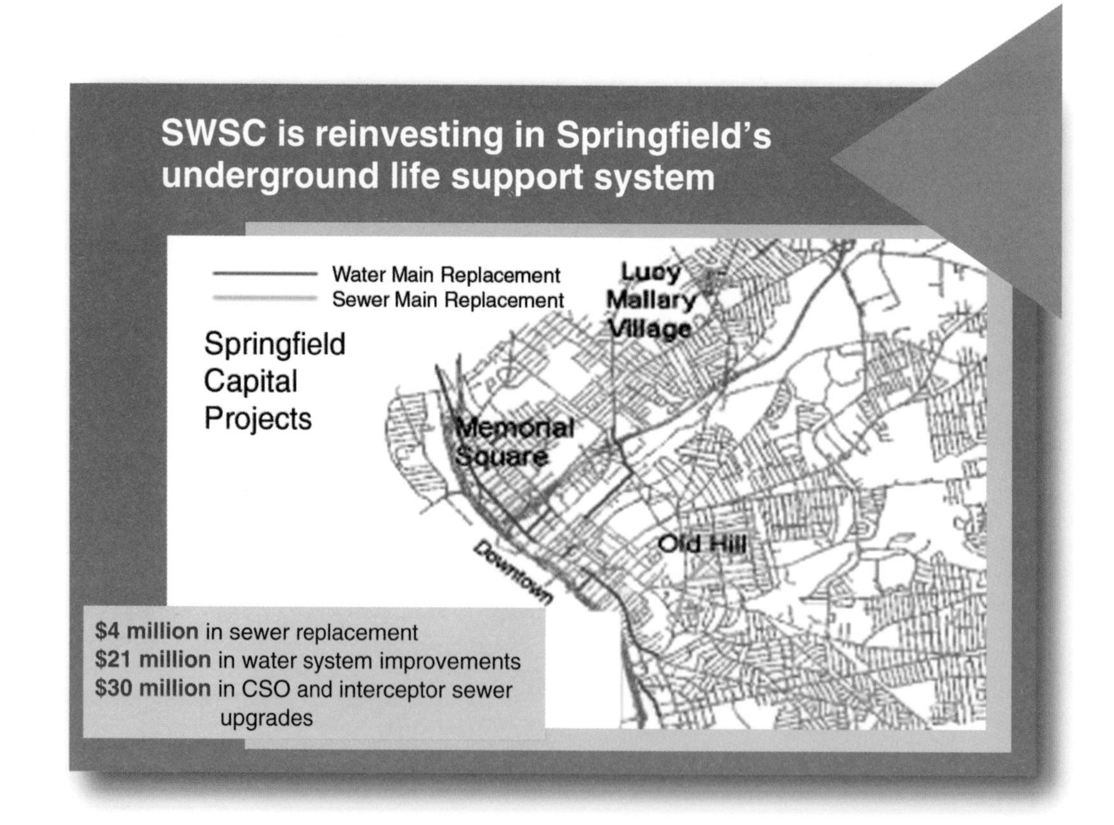

Improving our quality of life and leaving a legacy for future generations

- Controlling sewage overflows
- Wastewater treatment plant improvements
- Public outreach programs

Clean, uninterrupted water supply is a major asset

- Drinking water meets or surpasses all state and federal standards
- Protected from contamination—accidental or deliberate
- Tastes great too!

Securing a critical homeland asset—our water supply

- Tightened perimeter security
- Increased patrols
- Employee training
- Installing monitoring equipment
- Community watch programs

Capital budget: How SWSC invests for the future

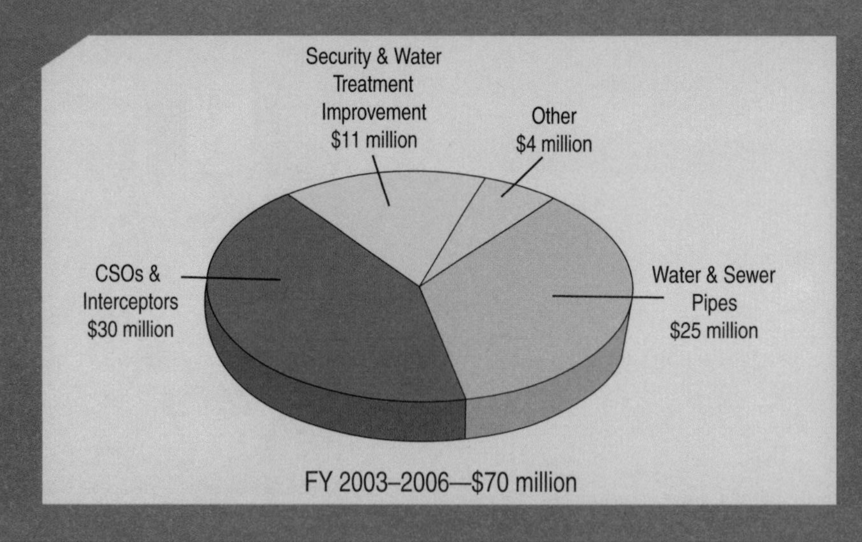

Additional costs due to post 9/11 security issues

- Chemicals (e.g. chlorine)
- Additional testing protocols
- Remote monitoring systems
- 24-hour shift changes
- Guard service
- Miscellaneous security measures (doors, locks, gates)
- Specialized training
- Increased insurance

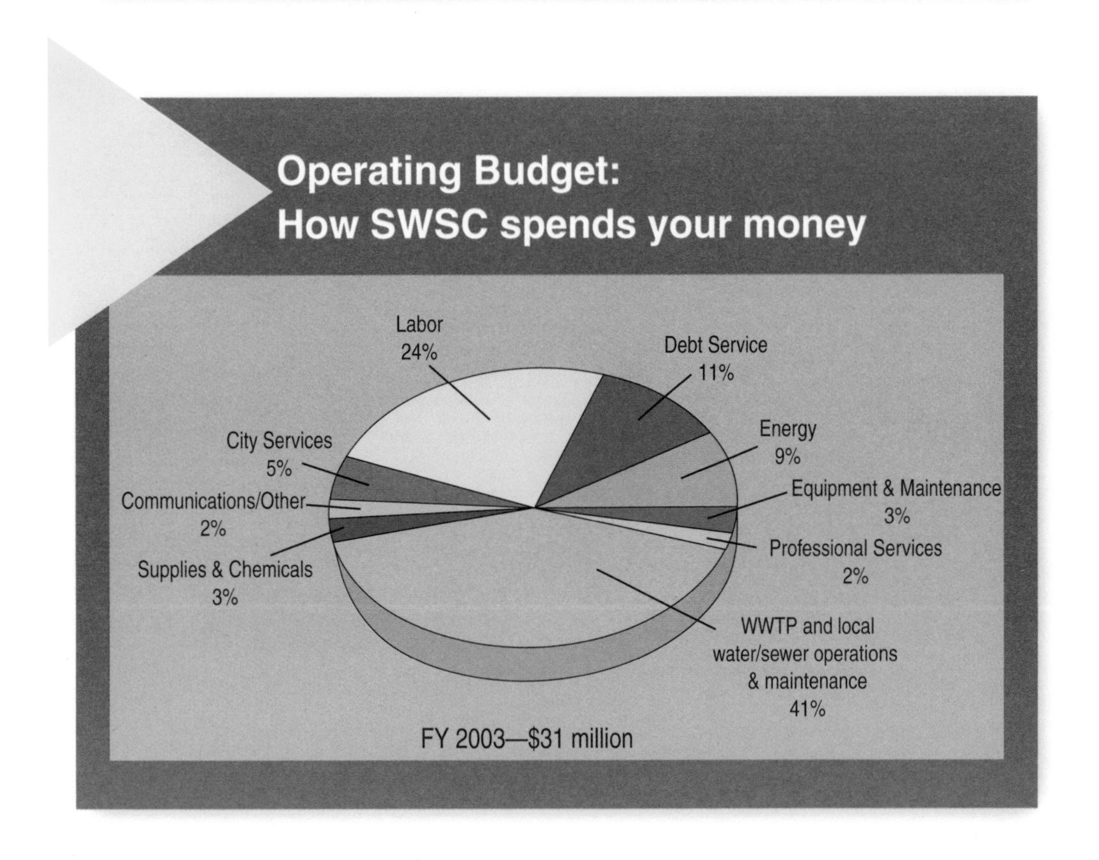

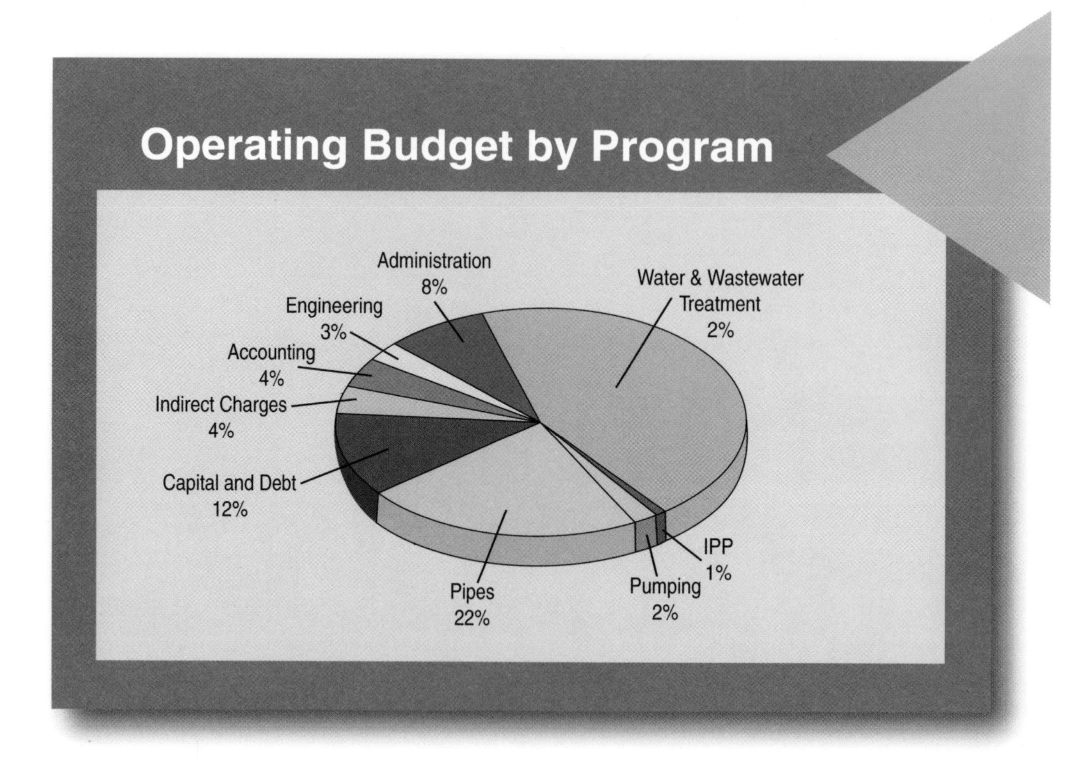
Operating Budget by Program
Administration
8%
Engineering
3%
Accounting
4%
Indirect Charges
4%
Capital and Debt
12%
Water & Wastewater
Treatment
2%
Pipes
22%
Pumping
2%
IPP
1%

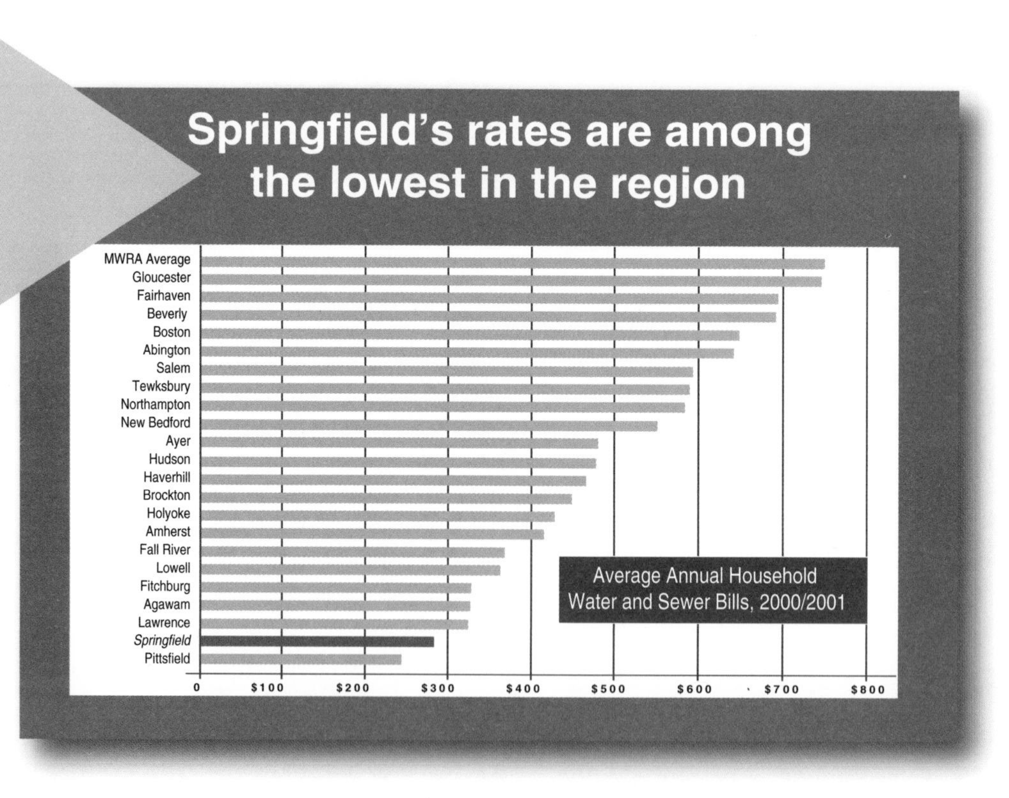
Springfield's rates are among
the lowest in the region
MWRA Average
Gloucester
Fairhaven
Beverly
Boston
Abington
Salem
Tewksbury
Northampton
New Bedford
Ayer
Hudson
Haverhill
Brockton
Holyoke
Amherst
Fall River
Lowell
Fitchburg
Agawam
Lawrence
Springfield
Pittsfield
Average Annual Household
Water and Sewer Bills, 2000/2001
0
$100
$200
$300
$400
$500
$600
$700
$800

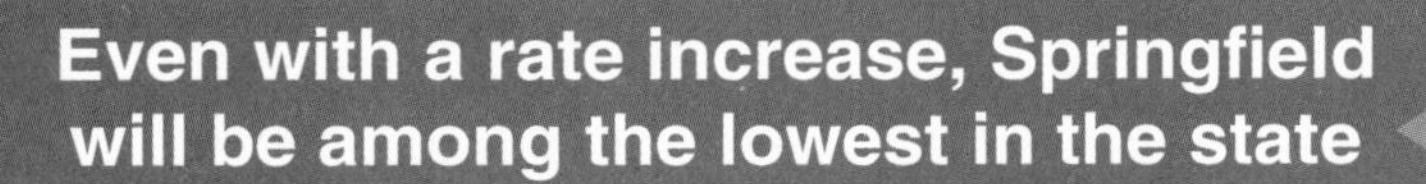
Even with a rate increase, Springfield will be among the lowest in the state

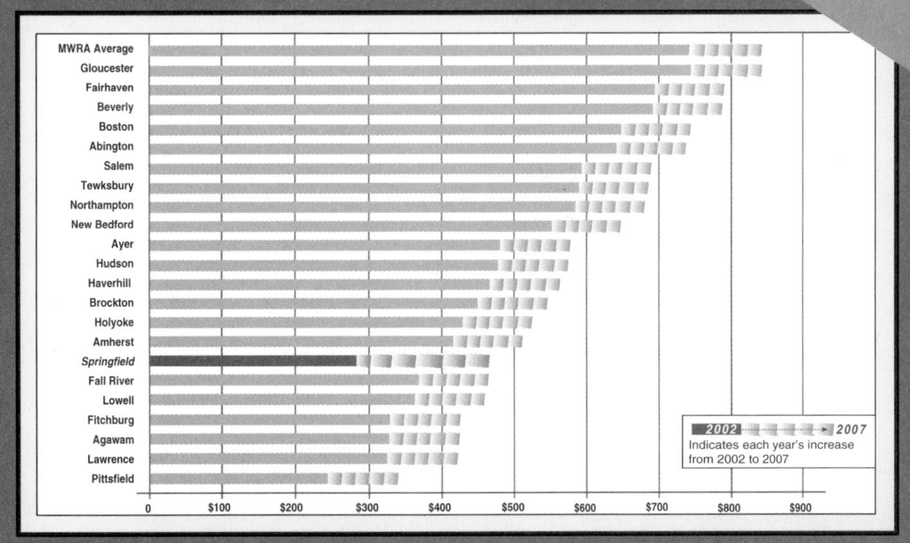
MWRA Average
Gloucester
Fairhaven
Beverly
Boston
Abington
Salem
Tewksbury
Northampton
New Bedford
Ayer
Hudson
Haverhill
Brockton
Holyoke
Amherst
Springfield
Fall River
Lowell
Fitchburg
Agawam
Lawrence
Pittsfield
0
$100
$200
$300
$400
$500
$600
$700
$800
$900
2002
2007
Indicates each year's increase from 2002 to 2007

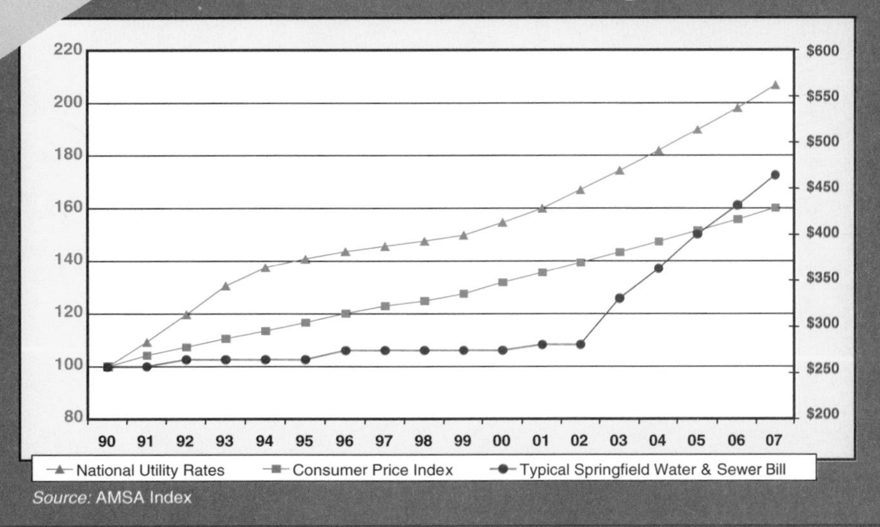
Springfield's rates have not kept pace with inflation and are significantly below the national average
220
200
180
160
140
120
100
80
$600
$550
$500
$450
$400
$350
$300
$250
$200
90 91 92 93 94 95 96 97 98 99 00 01 02 03 04 05 06 07
National Utility Rates
Consumer Price Index
Typical Springfield Water & Sewer Bill
Source: AMSA Index

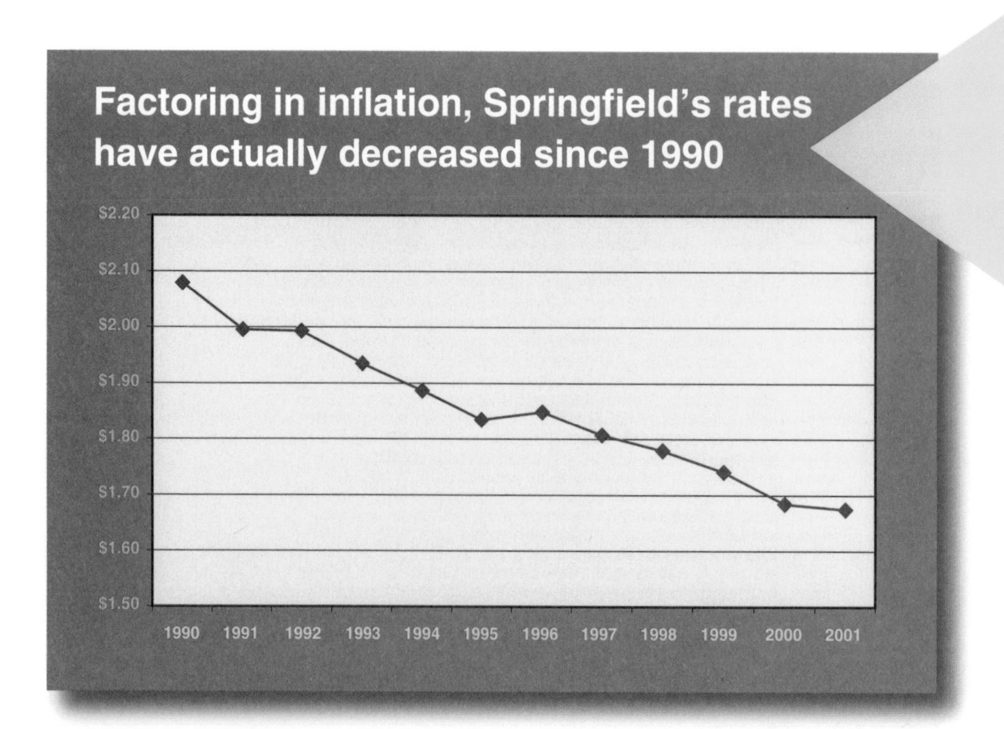

Water and sewer rates must rise to address increased costs

- Average annual increase of 11% over five years
- Typical household bill to increase $36.65 annually, or $3.05 per month
- Set water and sewer service charge replaces minimum water charge and administrative sewer charge

Special programs acknowledge different customer groups

- Elderly allowance of $30 per year
- Industrial rate structure encourages economic development
- Auxiliary meters allow for more accurate accounting of wastewater system use

Even at increased rates, Springfield's water is a bargain

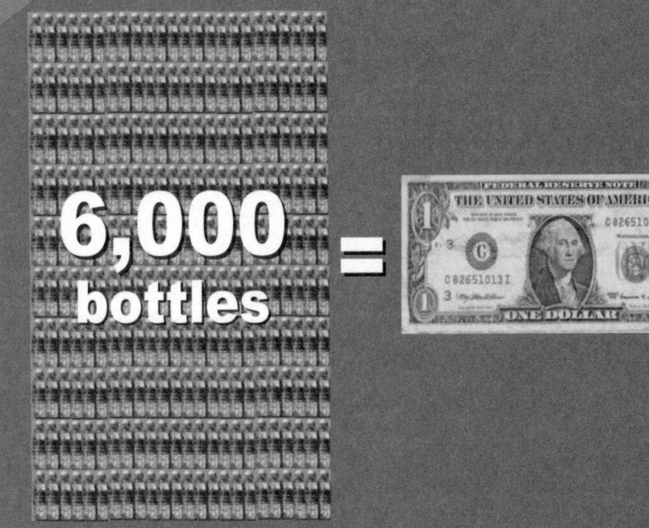

Even with constrained revenues, SWSC has delivered

- National awards for wastewater treatment
- Regional awards for Consumer Confidence Report
- Continuous compliance with the Safe Drinking Water Act
- Met or surpassed wastewater discharge permit requirements and eliminated odor problems from Bondi's Island plant

SWSC is working hard to make every dollar count

- Reorganization and public/private partnership at wastewater plant saved $5 million a year in avoided costs
- Meter replacement program provides automatic, accurate readings without entry by readers
- New financial system will improve accountability
- Monthly billing will provide up-to-date information for customer budgeting and water use management
- Collection program provides accountability and fairness for all ratepayers
- Water conservation programs help preserve limited supplies and decrease individual bills

APPENDIX C—AWWA POLICY STATEMENTS

Financing and Rates

The American Water Works Association (AWWA) believes the public can best be provided water service by self-sustained enterprises adequately financed with rates based on sound accounting, engineering, financial, and economic principles.

To this end, AWWA recognizes the following principles toward which water utilities should strive. Implementation of these principles should be balanced against other policy objectives; however, no policies should be adopted that compromise the long-term financial integrity of the water utility or its ability to provide service to customers. Basic financing and rate principles include:

1. Water utilities should receive sufficient revenues from water service, user charges, and capital charges, such as system development charges, to enable them to finance all operating and maintenance expenses and all capital costs (e.g., debt service payments).
2. Water utilities should account for and maintain their funds in separate accounts. Such funds should not be diverted to uses unrelated to water utility services. Reasonable payments in lieu of taxes and payments for services rendered to the water utility may be included in the cost of providing water service after taking into account the contribution for fire protection and other services furnished by the utility to local governments.
3. Water utilities should adopt a uniform system of accounts based on generally accepted accounting practices. The system of accounts should generally follow the accounting procedures outlined in the water utility accounting textbook published by AWWA. Modifications may be made to satisfy the financial needs of the utility and to meet the requirements of legislative, judicial, or regulatory bodies.
4. Water rate schedules should be designed to distribute the cost of water service equitably among each type of service and should reflect an appropriate balance of goals and objectives essential to the public good.

Asset Management

AWWA believes that water utilities must adopt a proactive approach to the management of their assets, which commences with planning and design and continues through operation and maintenance on to rehabilitation and replacement.

The purpose of water utility infrastructure is to provide costeffective, reliable supply and delivery of an adequate quantity of safe water. Each component of the infrastructure should be planned, designed and specified by professionals certified, licensed or accredited to perform the required task. Installation of each component should be performed by competent, trained, certified utility staff or licensed contractors. The complete system should be maintained by adequately trained personnel to ensure satisfactory performance. Satisfactory performance should be defined based upon the level of service requested by customers. When the continued performance of a component cannot be assured cost-effectively through prescribed maintenance, it should be rehabilitated or replaced using qualified professionals, trained utility staff, and/or licensed contractors as appropriate for the conditions.

An asset management plan for the operation and maintenance of each component should be established prior to its commissioning. Monitoring of this plan should be part of the management plan for the utility. Criteria that will signal the need for rehabilitation or replacement should be identified in the asset management plan and monitored during the operation and maintenance stages. Monitoring of operation and maintenance plans and ongoing component condition assessment should be included in the asset management plan for the utility. The plan should also address how renewal and replacement costs will be funded.

APPENDIX D

CREDIT IMPLICATIONS OF RATE STRUCTURE AND RATE SETTING FOR U.S. MUNICIPAL WATER-SEWER UTILITIES

—James Wiemken, Director, Standard & Poors Credit Market Services, January 20, 2004

As large future capital requirements for many water-sewer utilities become more apparent, whether because of regulatory compliance issues, sprawling growth, aging infrastructure, or simply better awareness of the need for long-term asset management, many utilities face an uphill battle to pay for these improvements due to political pressures and a lack of public understanding. Accordingly, utility credit analysis has moved beyond a point-in-time analysis of current debt service coverage and rates compared to ratepayer's income levels and rates in neighboring communities. Because a variety of factors may affect financing options at the local level, the extent of a utility's ability to implement strategies and policies that address its unique characteristics and allow it to finance needed projects becomes a differentiating factor. Many of the highest credit quality utilities rated by Standard & Poors also face significant capital requirements. The approaches they have taken, however, have allowed them to address these needs without sacrificing bondholder protection and without being hindered by political or public opposition. This article examines the key factors that Standard & Poors considers in relation to rate setting and capital planning when rating water-sewer bonds and uses several examples of utilities around the country which have made great progress in financing infrastructure to date while preparing for additional needs going forward.

Standard & Poors' analysis of rate setting practices centers on the question of whether rates are set such that available revenues are consistently sufficient to meet all of the ongoing needs and obligations of the utility, both now and in the future. While a variety of external factors influence this analysis, including regulatory issues, growth trends, customer concentration, and operational capacity, S&P generally looks for rate stability, rate transparency, and long-term planning as relevant factors that are under some control of utility management. Rate setting procedures that address these issues should help to achieve higher debt ratings, holding other factors constant.

Rate Stability Means Recognizing and Addressing Change

Achieving rate stability requires understanding that (more often than not) the statistic to be managed is the variance of the *changes* in rates over time rather than the variance in the rates themselves. Holding rate levels constant for multiple years does not benefit ratepayers if inflationary increases in operating costs and other expense pressures eventually compound to force a rate increase of such magnitude that rate payers have extreme difficulty in budgeting for this expense. Such patterns of irregular rate increases increase the risk that ratepayers will pressure rate makers to resist needed changes, thus increasing credit risk to bondholders. This is not to say that minimizing any negative economic development consequences of rate increases and pursuit of lower rates from further efficiencies should be ignored; they should be goals that are judged from a long-term perspective rather than exclusive targets to be met in the current year regardless of long-term consequences. When managed from a long-term perspective, sound policies usually benefit both bondholders and ratepayers, and the interests of these two constituencies are more consistently aligned.

Even without large financial pressures, political forces may intervene to delay rate increases, forcing the need for larger increases in the future. Unfortunately, this practice is somewhat common during election years for local officials. Such pressure is most damaging, however, when it continues for a multi-year period—often through the entire term of an administration. After several years of neglect, the local service area may face not only a current structural imbalance in its utility operations, but also compounded deferred maintenance, and a realization that the utility is no longer capable of addressing regulatory or growth related issues in a timely or manageable way. Unless credit ratings, auditor opinions, or other reports revealing this neglect receive attention, these practices may continue for several years because many systems have funds set aside for improvements which (in some cases) may be diverted for rate subsidization.

Some utilities have created rate stabilization funds that technically exist to smooth rate increases over a long-term period. While Standard & Poors' generally prefers that rates be regularly set to provide sufficient funds to meet current obligations, such funds can be credit strengths when used appropriately. Whereas the use of capital and other funds for avoiding needed rate increases detracts from the utility's long-term stability, the use of specifically designated rate stabilization funds (generally resulting from surplus moneys) to reduce (but not eliminate) an atypically large rate increase, may benefit a utility. For example, although Boston Water and Sewer Commission will need to continue to raise rates following 8.9% increases at the beginning of both 2002 and 2003 and an additional 3.9% in April 2003, it will use over $20 million of an available $47 million in stabilization funds going forward to keep increases manageable. If the rate stabilization fund use is not a recurring reliance and it gives the utility more credibility in achieving needed rate increases going forward, this practice can be credit strength. It should be noted, however, that rate makers may use rate stabilization funds for political convenience, and the temptation to rely on further subsidization from the fund to meet ongoing expenditure pressures may be great. Even with a fund specifically designated for rate stabilization and funded from surplus moneys, this practice would be considered a credit risk as one-time funds were being used to temporarily meet a long-term expenditure. Without an identified long-term revenue source to make up for the rate subsidization in the following year, the utility's financial structure would be positioned to weaken consistently going forward. S&P stresses the importance of rates producing current revenues to meet current obligations for this reason.

When future financial needs are known, multi-year rate approvals are another tool to protect rate setting from outside influence. If rates are regularly adjusted, then incremental increases should be smaller and thus easier to approve—even as a whole. Approving rates for multiple years at a time often allows a utility to lock in funding for the entire cost of a needed improvement with one effort, rather than having to re-explain the reasons for the project each time a new phase of the rate increase is needed. While statutes and oversight provisions limit this ability in some states, even internal policies or agreements in principle can help build and maintain support for needed increases through politically sensitive periods.

Rate Transparency

Ratepayers will often accept rate increases when they are manageable, but understanding the reasons for needed rate increases becomes more important when larger increases are necessary. While good communication with ratepayers and all stakeholders is generally the best prescription for transparency, many utilities structure their rate setting policies so as to ensure a certain level of transparency. Many local water utilities purchase most if not all of their water pre-treated from another water utility on a wholesale basis. These distribution utilities generally face fewer regulatory burdens and water purchases from the wholesaler may constitute the vast majority of its expenses. Policies which automatically pass through any wholesale rate increases to retail customers allow ratepayers to equate their rate increases with the timing of the wholesale increases, thus creating better understanding.

By communicating the frequency and degree to which retail increases result from wholesale increases, the retail provider may enjoy better support for non-wholesale related increases when the need arises. While insulating retail customers from wholesale increases may be politically popular in the short term, it can have devastating consequences for the long term if an additional long-term savings or another funding source cannot be identified to meet this expenditure pressure.

Anaheim, California takes cost transparency one step further by breaking its water charge into a base charge and a commodity charge, which includes the cost of purchased water and the cost of electricity. Ratepayers can therefore better understand the degree to which their rates are influenced by short- or long-term factors. The use of a base charge can also be a credit strength in that funds anticipated for debt service payments are subject only to fluctuations in the number of customers—not to fluctuations in consumption.

While many utilities focus on the transparency of cost pressures because they most often drive the need for rate increases, the transparency of benefits should not be neglected. Occasionally rate increases are needed for improvements that lead to observable benefits in quality (such as removing taste and odor), but customers may be unaware of the relative benefits they are receiving. Cincinnati, Ohio's water utility has benefited greatly from its reputation for high-quality water. Over time, the utility has grown its service area as neighboring residents actively sought access to this resource. The demand from residents in northern Kentucky was such that the area utility actually chose to tunnel underneath the Ohio River to hook up to the Cincinnati system rather than developing its own treatment systems (cost played a role as well). Such public recognition makes for good relations not only with new customers who have not always enjoyed access to Cincinnati's water, but also with lifelong customers. To date, the city council has never failed to approve a rate increase proposed by the utility.

For those utilities whose water quality may not differ substantially from its neighbors, other performance measures are often used, including rate increases vs. inflation, cost increases vs. inflation, total employees, total customers per employee, and a variety of other services delivered vs. cost measures. Although such measures must often be explained and clarified when system changes occur, encouraging ratepayers to focus on the marginal or relative benefits they receive is helpful in looking beyond a long-term trend of rising rates.

Long Range Planning

Policies encouraging rate stability and transparency over the short- and medium-term horizons may be implemented with some success, but they are likely to prove insufficient without some focus on relating the system's current status to its long-term needs. True rate stability and transparency assumes that a system's current and likely future needs have been measured and are relatively known. The average increase in rates to be targeted over the next decade cannot be known without some idea of the cost pressures a utility may face, and without an honest effort to estimate these needs, it will be extremely difficult to educate and inform ratepayers. Cost pressures to be estimated include those for operations, replacement, regulatory compliance, and accommodating additional growth. The nature of these cost increases should be considered (i.e. whether they are ongoing or likely to be diminished over time) along with their magnitude. Opportunities for savings should also be considered, which could result from technological improvements or administrative restructuring. The components of the revenue stream should also be examined. How much revenue is coming from connection fees and other one-time sources, and how this relates to current and expected growth trends is especially important. If the utility relies on a single commercial or industrial customer, the likelihood of that entity maintaining its current presence in the service area over the next ten to twenty years should be considered, as well as how reliant the utility wants to be on this assumption.

Many utility officials site the impossibility of correctly estimating future economic development trends, regulatory outcomes, and the long-term patterns of various cost pressures. As such, they claim that trying to measure them actually represents a poor use of limited resources, especially for smaller systems that lack the staff or funds for consultants to devote to such studies. While most of these drivers are indeed highly uncertain, Standard & Poors' views a refusal to consider the potential burden of pressures beyond the short to medium term to be a credit risk. Accordingly, even small utilities that have attempted to examine long-term risks and possibilities in limited ways consistent with their resources and capabilities will likely find their rate projections and capital plans more accepted by S&P.

Conclusion

Municipal Water and Sewer utilities are forced to address a variety of short-term and long-term pressures on a regular basis. While a system's current financial status is of some importance to the utility's credit rating, its likely long-term health is the key driver. As the likelihood of significant additional capital needs increases, the current rate, financial, and debt pictures for a utility become less reliable as indicators of long-term credit quality in and of themselves. A utility's ability to implement policies and procedures which garner the support of ratepayers for the additional revenues required to support these needs will become more important to the rating. Such policies should encourage both rate stability and transparency, and should minimize the likelihood of political influence that sacrifices the utility's long-term health for temporary rate freezes. While the exact nature of the future challenges and demands on a utility is impossible to forecast, early efforts to plan for long-term pressures will allow the utility to address the needs over a longer time horizon and in a more manageable way that benefits both ratepayers and bondholders.

For more information, please contact:

James Wiemken, Director
Standard & Poors Credit Market Services
One Prudential Plaza
130 East Randolph St., Suite 2900
Chicago, IL 60601
312.233.7005 Tel
312.233.7051 Fax
james_wiemken@sandp.com

REFERENCES

Akron Beacon Journal, "Averting blackout crisis: Cleveland to invest in generators to keep water flowing in future," November 15, 2003.

AWWA, *Dawn of the Replacement Era*, May 2001.

AWWA, *Principles of Water Rates, Fees, and Charges:* AWWA Manual M1—Fifth Edition, 2000.

AWWA/National Consumer Law Center, *Water Affordability Programs,* 1998.

AwwaRF/CH2MHILL, *Public Involvement Strategies: A Manager's Handbook,* 1995.

CH2M HILL/*City of Syracuse, New York, City of Syracuse Water Study Summary*, November 2003.

Congressional Budget Office, *Future Investment in Drinking Water and Wastewater Infrastructure*, November 2002.

Hoffbuhr, Jack, "Was Malthus Right?" *Journal AWWA*, p. 6, August 2003.

Hurd, Robert, *Consumer Attitude Survey on Water Quality Issues,* AwwaRF Report #90654, 1993.

Pew Internet & American Life Project/*Federal Computer Week* Magazine, "Internet and Emergency Preparedness Survey," August 2001. http://www.pewinternet.org/reports/toc.asp?Report=100

Public Policy Institute of California/University of California, Irvine, *2002 Statewide Survey:* December 2002. http://data.lib.uci.edu/ocs/2002/report/02infrast.html

San Diego County Water Authority, *Telephone Public Opinion Survey,* 2003. http://www.sdcwa.org/about/pdf/2003_SurveyReport.pdf

Speranza, Elisa, "Death of the Silent Service: Meeting Customer Expectations," Chapter 17, *Drinking Water Regulation and Health,* edited by Frederick Pontius, John Wiley & Sons, Inc. Publishers, 2003.

Tampa Water Department, *Customer Service Survey,* May 2003. http://www.tampagov.net/dept_water/conservation_education/pdf/Water%20Survey%20Report.pdf

University of Idaho and U.S. EPA Region 10, *Survey of Public Attitudes about Water Issues in the Pacific Northwest*, 2002. http://yosemite.epa.gov/R10/ecocomm.nsf/34090d07b77d50bd88256b79006529e8/55d34260c046dc7e88256c2f00565985

U.S. EPA, *Analysis and Findings of the Gallup Organization's Drinking Water Customer Satisfaction Survey,* 2003.

U.S. EPA, "Sustainable Water Infrastructure for the 21st Century," http://www.epa.gov/water/infrastructure/index.htm.

Wiemken, James, Director, Standard & Poors Credit Market Services, "Credit Implications of Rate Structure and Rate Setting for U.S. Municipal Water-Sewer Utilities," white paper, January 20, 2004.

Headquarters Office

6666 West Quincy Avenue
Denver, CO 80235
303.794.7711

Fax: 303.794.1440

http://www.awwa.org

Government Affairs Office

1401 New York Avenue
NW, Suite 640
Washington, DC 20005

202.628.8303

Fax 202.628.2846